DATE

INORGANIC TITRIMETRIC ANALYSIS

CONTEMPORARY METHODS

TREATISE ON TITRIMETRY

Editor in Chief
JOSEPH JORDAN
Department of Chemistry
The Pennsylvania State University
University Park, Pennsylvania

INORGANIC TITRIMETRIC ANALYSIS

Contemporary Methods

Walter Wagner and Clarence J. Hull

University of Detroit
Detroit, Michigan

Indiana State University
Terre Haute, Indiana

MARCEL DEKKER, INC. NEW YORK   1971

MARCEL DEKKER, INC.
95 Madison Avenue, New York, New York 10016

LIBRARY OF CONGRESS CATALOG CARD NUMBER: 70-160515
ISBN NO.: 0-8247-1743-0

PRINTED IN THE UNITED STATES OF AMERICA

PREFACE

This is an unusual book. Its format is predicated by a very specific purpose. It is intended to serve as a rapid and convenient guide in situations frequently encountered by all practicing chemists. The need which this volume is intended to meet involves the following typical predicament. A method is required for determining quantitatively an element (or a simple inorganic compound), e.g., fluoride in a kinetic study, silica in a process stream, mercury in water, sulfur dioxide in polluted air. The requirement is for limited duration, i.e., it is either a "one shot" affair to make a single "spot check," or involves a relatively small number of analyses in a pilot project. Under these conditions it is imperative that the analysis be feasible with the facilities available in a well-equipped average contemporary chemical research or service laboratory, which normally include potentiometric, coulometric, and spectrometric equipment but not necessarily a high flux nuclear reactor. In a compromise between the unwieldy cumbersomeness of classical gravimetric procedures on the one hand, and the need for ad hoc equipment in using highly sophisticated instrumental methods on the other hand, a preferred choice is modern titrimetry oftentimes combined with an instrumental end point (amperometric, spectrophotometric, thermometric, etc.). In this context, it is fair to say that no chemist on earth is likely to "get by" without using, sometimes, some form of titrimetric analysis. Indeed while the day-to-day use of gravimetric analysis has been steadily declining, titrimetry continues to remain an important tool in diversified areas of quantitative chemistry, including industrial process control, environmental pollution monitoring, and clinical laboratories. Titrimetric analysis has been significantly enriched by several contemporary developments, such as multidentate chelating reagents (e.g., EDTA), or ingenious coulometric generation of redox titrants in situ.

The chemist seeking a convenient titrimetric method has had in recent years a frustrating choice among several unsatisfactory alternatives, viz.: rely on a specialized reference book of volumetric analysis, the newest of which dates back to the fifties; use "handbooks" which are concise but sketchy; "plow through" multivolume general treatises of analytical chemistry; or make an exhaustive and time-consuming search of cumulative indexes of journals (e.g., ANALYTICAL CHEMISTRY or CHEMICAL ABSTRACTS). The object of this book is to provide a suitable "shortcut," circumventing these difficulties. The authors have set themselves the task of compiling, in one single normal-sized volume, judiciously selected information on inorganic titrimetric analysis. Selectivity was exercised along the following guidelines:

1. Coverage has been restricted to material which--in the opinion of the authors--is likely to be used in chemical

iii

research and industry in this day and age: outdated methods
have been omitted altogether.
2.   There is no pretense of exhaustive literature referencing.
Entries in bibliographies given at the end of sections (each
devoted to an element) have been restricted to ten (10) or
less. The bulk of the bibliography consists of material
published since 1960.
3.   The authors have exercised discretion in summarizing concisely
a couple of "recommended" contemporary methods for each element.

The philosophy of this book is to provide sufficient information
for readers to decide whether a given method of analysis might suit
their needs, in terms of applicability, specificity, precision, accuracy, proneness to interference, as well as in terms of reagents and
instruments required. Whenever possible, a choice of several methodological approaches is offered (e.g. acid-base, precipitation, complexation and/or redox titrations): advantages and limitations are
pointed out, to enable the reader to select the specific methods which
may be best suited for his problem. However, no attempt has been made
to include exhaustive details of procedure and technique. For this
the reader is advised to consult the pertinent literature.

To implement the philosophy, sections devoted to each element
(and its inorganic compounds) have been organized along an identical
plan, which consists of:

a.   a concise statement of the general types of methodological
approaches available;
b.   a synoptic survey of applicable "classical" (pre 1950) and
"contemporary" (mostly post 1960) titrimetric methods;
c.   a brief description and characterization of a few "recommended contemporary methods";
d.   a bibliography restricted to judiciously selected references.

The emphasis in this book is on the pragmatic side. It is deliberately selective and deliberately incomplete in order to be useful.
It is not an exercise in creative writing, but endeavors to serve an
audience ranging from graduate students to laboratory directors and
practicing chemists engaged in research, development, analysis, manufacture, process control, etc.

Salient features of this book include sections on such novel
developments as redox titrimetry of the noble gas xenon and the determination of transuranium elements. Care has been taken to cover ade-
quately significant contributions emanating from Russia and other East
European countries, which are not readily accessible in English else-
where. Last but not least, this book has no index. Chapters are
arranged according to the sequence of the elements in the periodic
table which obviates any further subject indexing. An author's index
was omitted advisedly: a Pantheon of this type would serve no purpose
in a book whose emphasis is on usefulness.

The present volume is part of a treatise on TITRIMETRY.

Companion Volumes on "CONTEMPORARY METHODS OF ORGANIC FUNCTIONAL GROUP ANALYSIS", "PRINCIPLES OF TITRIMETRY", and "FIELDS OF APPLICATION" will be published separately.

University Park, Pennsylvania   Joseph Jordan
June, 1971   Editor in Chief

PERIODIC TABLE OF THE ELEMENTS

```
                                                                        GROUP
```

atomic number → 1
H
1.0080 ← atomic weight (see footnote)

TRANSITION ELEMENTS

IA	IIA	IIIB	IVB	VB	VIB	VIIB	VIII			IB	IIB	IIIA	IVA	VA	VIA	VIIA	O
1 H 1.0080																	2 He 4.003
3 Li 6.939	4 Be 9.012											5 B 10.81	6 C 12.011	7 N 14.007	8 O 15.9994	9 F 18.998	10 Ne 20.183
11 Na 22.990	12 Mg 24.31											13 Al 26.98	14 Si 28.09	15 P 30.974	16 S 32.064	17 Cl 35.453	18 Ar 39.948
19 K 39.102	20 Ca 40.08	21 Sc 44.96	22 Ti 47.90	23 V 50.94	24 Cr 52.00	25 Mn 54.94	26 Fe 55.85	27 Co 58.93	28 Ni 58.71	29 Cu 63.54	30 Zn 65.37	31 Ga 69.72	32 Ge 72.59	33 As 74.92	34 Se 78.96	35 Br 79.909	36 Kr 83.80
37 Rb 85.47	38 Sr 87.62	39 Y 88.91	40 Zr 91.22	41 Nb 92.91	42 Mo 95.94	43 Tc (99)	44 Ru 101.1	45 Rh 102.90	46 Pd 106.4	47 Ag 107.870	48 Cd 112.40	49 In 114.82	50 Sn 118.69	51 Sb 121.75	52 Te 127.60	53 I 126.90	54 Xe 131.30
55 Cs 132.91	56 Ba 137.34	57 to 71 La 138.91	72 Hf 178.49	73 Ta 180.95	74 W 183.85	75 Re 186.2	76 Os 190.2	77 Ir 192.2	78 Pt 195.09	79 Au 197.0	80 Hg 200.59	81 Tl 204.37	82 Pb 207.19	83 Bi 208.98	84 Po (210)	85 At (210)	86 Rn (222)
87 Fr (223)	88 Ra 226.05	89 to 103															

INNER TRANSITION ELEMENTS

Lanthanide series	57 La 138.91	58 Ce 140.12	59 Pr 140.91	60 Nd 144.24	61 Pm (147)	62 Sm 150.35	63 Eu 151.96	64 Gd 157.25	65 Tb 158.92	66 Dy 162.50	67 Ho 164.93	68 Er 167.26	69 Tm 168.93	70 Yb 173.04	71 Lu 174.97
Actinide series	89 Ac (227)	90 Th 232.04	91 Pa (231)	92 U 238.03	93 Np (237)	94 Pu (242)	95 Am (243)	96 Cm (247)	97 Bk (249)	98 Cf (251)	99 Es (254)	100 Fm (253)	101 Md (256)	102 No (254)	103 Lw (257)

$C^{12} = 12.00000$; a value in parenthesis for the atomic weight indicates the mass number of longest known half life

ALPHABETICAL LISTING OF ELEMENTS COVERED

Element	Atomic Number	Element	Atomic Number	Element	Atomic Number
Aluminum	13	Iodine	53	Rubidium	37
Antimony	51	Iridium	77	Ruthenium	44
Arsenic	33	Iron	26	Scandium	21
Barium	56	Lanthanum	57	Selenium	34
Beryllium	4	Lead	82	Silicon	14
Bismuth	83	Lithium	3	Silver	47
Boron	5	Lutetium	71	Sodium	11
Bromine	35	Magnesium	12	Strontium	38
Cadmium	48	Manganese	25	Sulfur	16
Calcium	20	Mercury	80	Tantalum	73
Carbon	6	Molybdenum	42	Technetium	43
Cerium	58	Neodymium	60	Tellurium	52
Cesium	55	Neptunium	93	Terbium	65
Chlorine	17	Nickel	28	Thallium	81
Chromium		Niobium	41	Thorium	90
Cobalt	27	Nitrogen	7	Thulium	69
Copper	29	Osmium	76	Tin	50
Dysprosium	66	Oxygen	8	Titanium	22
Europium	63	Palladium	46	Tungsten	74
Fluorine	9	Phosphorus	15	Uranium	92
Gadolinium	64	Platinum	78	Vanadium	23
Gallium	31	Plutonium	94	Xenon	54
Germanium	32	Polonium	84	Ytterbium	70
Gold	79	Potassium	19	Yttrium	39
Hafnium	72	Rhenium	75	Zinc	30
Hydrogen	1	Rhodium	45	Zirconium	40
Indium	49				

CONTENTS

Contents

Group O

XENON

The only recommended procedure for the determination of xenon involves redox titration.

Synoptic Survey

Classical Methods

None

Contemporary Methods

Redox Titration. Xenates and perxenates are analyzed iodometrically. The "high" value, Xe(VIII) + Xe(VI), is determined by adding excess potassium iodide followed by acidification with sulfuric acid [1]. The "low" value, Xe(VI), is determined by first acidifying with sulfuric acid followed by the addition of excess potassium iodide.

In each case, the liberated iodine is titrated with standard thiosulfate to an amylose end point [2].

Outline of Recommended Method for Xenon

Redox Titration

In acidic solutions iodide reduces Xe(VI) to elemental Xe. In an alkaline medium, the reaction of Xe(VI) with iodide is very slow. The Xe(VIII) reacts with iodide in both basic and acidic media. In an acidic solution, Xe(VIII) is reduced to Xe(VI) accompanied by the liberation of oxygen. The oxidizing power of a xenon salt is determined by adding excess potassium iodide, acidifying with sulfuric acid, and titrating the triiodide ion so formed with standard thiosulfate (which has been standardized against potassium permanganate) to an amylose end point:

$$Xe(VIII) + 12\ I^- \rightarrow Xe° + 4\ I_3^-$$

$$Xe(VI) + 9\ I^- \rightarrow Xe° + 3\ I_3^-$$

If the solution is acidified before the iodide is added, only Xe(VI) is found since any octavalent xenon would be decomposed to Xe(VI) and oxygen:

1

$$Xe(VIII) \quad \xrightarrow{\;\;H+\;\;} \quad Xe(VI) + O_2$$

$$Xe(VI) + 9 \; I^- \;\rightarrow\; Xe^\circ + 3 \; I_3^-$$

Since the "oxidizing power" will differ depending on whether $Xe(VI)$ and/or $Xe(VIII)$ are present, the ratio of $Xe(VI)$ to $Xe(VIII)$ can be determined.

References

[1]. Private communication from Dr. T. M. Spittler, and Sister M. Raphael Crumb, O.S.U., Univ. of Detroit, Detroit, Michigan.

[2]. E. H. Appelman, Noble-Gas Compounds, H. H. Hyman, ed., Univ. of Chicago Press, 1963, pp. 185-90.

GROUP I

Hydrogen

Lithium   Copper

Sodium   Silver

Potassium   Gold

Rubidium

Cesium

HYDROGEN

Titrimetric Procedures for the determination of water and hydrogen peroxide are outlined in group II as part of the discussion of oxygen.

LITHIUM

Titrimetric procedures for the determination of lithium include acid-base, redox, and precipitation methods.

Synoptic Survey

Classical Methods

Redox Titration. Lithium is precipitated as a complex periodate in strongly alkaline potassium periodate solution. The precipitate is titrated iodometrically with sodium thiosulfate or sodium arsenite [1,2].

Precipitation Titration. The Volhard method as applied to lithium involves the precipitation of lithium chloride followed by its separation from any sodium or potassium chloride by extraction with 2-ethyl-1-hexanol. The extracted lithium chloride is then titrated directly in the alcoholic phase after a single extraction [3].

Contemporary Methods

Acid-Base Titration. Lithium is determined by potentiometric

3

titration of the acetate with standard perchloric acid in acetic acid
solutions [4].

Precipitation Titration. Lithium hydroxide is determined by a
conductometric titration using orthophosphoric acid as the titrant [5].

Outline of Recommended Contemporary Methods for Lithium

Acid Base Titration

Lithium acetate is titrated potentiometrically with 0.1 M or 0.5
M perchloric acid in acetic acid. The solvent used is a 1:4 benzene-
nitrobenzene mixture modified with acetic acid containing 5% acetic
anhydride. Lithium salts must first be converted into the sulfate by
treatment with sulfuric acid, and then subsequently into the acetate
by reaction with barium acetate. The relative error is 0.20% in the
determination of the single salt. In the titration of ternary mixtures,
the precision in the determination of 0.25 meq amounts of lithium, sodium, and potassium is $\pm$ 0.039, $\pm$ 0.045, and $\pm$ 0.33 meq respectively
[4].

Precipitation Titration

Lithium is determined in the presence of sodium, potassium,
ammonium, and magnesium ions by converting the chlorides of these cations into their respective hydroxides by the addition of silver oxide
to an ammoniacal solution of the ions. The silver-ammonia complex is
then decomposed by dilution with ethyl alcohol; the silver chloride
and magnesium hydroxide are removed by filtration. The lithium is
finally precipitated as the phosphate by conductometric titration
using 0.1 N orthophosphoric acid. The titration is carried out in 70%
ethyl alcohol at 20°. The optimum ratio of lithium hydroxide to the
other hydroxides is 1:5. If such is not the case, and if the concentration of the total amount of the other hydroxides is less than 1:5,
then the optimum ratio of lithium hydroxide to the other hydroxides
is best achieved by the addition of 0.1 N sodium hydroxide dissolved
in 70% ethyl alcohol. The standard deviation is 1.10%. The method
is both simple and rapid [5].

References

[1]. L. B. Rogers and E. R. Caley, *Ind. Eng. Chem. Anal. Ed.*, *15*,
      209 (1943).

[2]. F. R. Bacon and D. T. Starks, *Ind. Eng. Chem. Anal. Ed.*, *17*,
      230 (1945).

[3]. J. E. White and G. Goldberg, *Anal. Chem.*, *27*, 1188-1189 (1955).

[4]. S. Kiciak, *Chem. Anal. (Warsaw)*, *12*(2), 209-216 (1967).

[5]. V. Novak and S. V. Dolejs, *Sb. Ved. Praci., Vysoka Skola Chem. Technol., Pardubice, 1963*(1), 39-42.

SODIUM

No direct method is available for the titrimetric determination of sodium. Indirect procedures involve acid-base and complexometric titrations.

Synoptic Survey

Classical Methods

Acid-Base Titration. Sodium is determined by the titration of a pyroantimonate solution with 0.1 N hydrochloric acid [1,2].

Redox Titration. Sodium is precipitated in 30% alcoholic solution as the pyroantimonate and titrated iodometrically [3].

Contemporary Methods

Acid-Base Titration. Sodium is precipitated as sodium hydrogen di(α-methoxy)phenylacetate and titrated with standard base [4].

Complexometric Titration. Sodium is precipitated as sodium zinc uranyl acetate hexahydrate. The precipitate is dissolved in hot aqueous ethanol and the zinc titrated with a standard solution of EDTA [5].

Outline of Recommended Contemporary Methods for Sodium

Acid-Base Titration

Sodium hydrogen di(α-methoxy)phenylacetate, $C_6H_5CH(OCH_3)COOH \cdot C_6H_5CH(OCH_3)COONa$, is sparingly soluble in cold water, but dissolves readily in hot. Thus sodium can be precipitated and determined by titration of the above acid salt with standard base. The method is highly selective since sodium is the only common metal precipitated in this manner. Potassium and lithium do not interfere. A standard solution of a strong base is used as the titrant in conjunction with phenolphthalein as the indicator. If ammonium, magnesium, or heavy metal salts are present, chlorophenol red is the preferred indicator. Using 0.05 N base as the titrant, the standard error is 5 parts per 1000 for samples containing 45 mg of sodium; however, the precipitate must be allowed to stand overnight at 0-2° before titration in order to insure complete precipitation. For more rapid analysis when time does not permit the sample to stand overnight, a correction is added to the amount of sodium found to allow for a constant loss due to slight solubility of the sodium acid salt [4].

Chelation Titration

Sodium is precipitated as sodium zinc uranyl acetate hexahydrate, $NaZn(UO_2)_3(CH_3COO)_9 \cdot 6H_2O$. The precipitate is filtered and washed, then dissolved in hot water and ethyl alcohol. The zinc is titrated with 0.001 M EDTA using dithizone as the indicator. During titration, the solution is buffered with sodium acetate-acetic acid. The amount of sodium which may be determined in this manner is 80-320γ. The accuracy is claimed to be "acceptable" [5].

References

[1]. W. Hurka, *Z. Physiol. Chem.*, *276*, 30-37 (1942).

[2]. W. Hurka, *Mikrochemie*, *30*, 297 (1942).

[3]. H. Mueller, *Helv. Chim. Acta*, *6*, 1156 (1924).

[4]. W. Reev, *Anal. Chem.*, *31*, 1066-1068 (1959).

[5]. A. Holasek and M. Dugandzic, *Mikrochim. Acta*, 1959, 488-489.

POTASSIUM

Titrimetric procedures for the determination of potassium include redox, complexometric, and precipitation methods.

Synoptic Survey

Classical Methods

None.

Contemporary Methods

Redox Titration. Potassium is precipitated with periodic acid as K_5IO_6. The precipitate is dissolved in sulfuric acid and titrated iodometrically with sodium thiosulfate [1].

Complexometric Titration. To a solution of potassium tetraphenylboron in N,N-dimethylformamide, an excess of mercury(II)-EDTA is added. The liberated EDTA is then back-titrated with magnesium solution [2].

Precipitation Titrations. Potassium in an acetic acid-chloroform medium is titrated with perchloric acid in dioxane [3].
Potassium tetraphenylboron is titrated coulometrically with electrogenerated silver ion in an aqueous-acetone media [4].
Potassium ion is titrated with tetraphenylborate ion. A cationic sensitive glass electrode serves as an indicator electrode [5].
Potassium ion is titrated potentiometrically with zinc hexa-

fluosilicate. The end point is detected with a special glass electrode [6].

Outline of Recommended Contemporary Methods for Potassium

Redox Titration

The sample containing no more than 75 mg of potassium and no
oxidizing agents, reducing agents, or halides, is precipitated with
periodic acid. In addition to the potassium, only lithium, sodium, or
the alkaline earth cations should be present. After precipitation, the
pH of the solution is adjusted to 3.0-3.5 using a lithium acetate buffer. The mixture is then allowed to stand for 30-40 minutes at $0°C$ before the KIO_4 precipitate is removed and dissolved in dilute H_2SO_4. An
excess of KI is then added to the acidic KIO_4 solution, and the liberated iodine is titrated with standard sodium thiosulfate using a starch
indicator [1].

Chelation Titration

Potassium is precipitated as the tetraphenylborate and the precipitate so obtained is dissolved in N,N-dimethylformamide. An excess
of mercury(II)-EDTA is added, and the liberated EDTA is back-titrated
with a standard magnesium solution according to the equation:

$$4HgY^{-2} + (C_6H_5)_4B^- + (4n)H_2O \rightarrow H_3BO_3 + 4C_6H_5Hg^+$$
$$+ 4H_nY^{n-4} + (4n-3)OH^-$$

$$[HgY^{-2} = mercury(II) \ chelonate]$$

Eriochrome Black T is used as the indicator. A single analysis takes
approximately 20 minutes. The standard deviation from the mean in
milligrams is $\pm$ 0.003 for 18.57 mg potassium; $\pm$ 0.028 for 4.889 mg
potassium; and $\pm$ 0.003 for 0.489 mg potassium [2].

Precipitation Titrations

Method 1. Potassium is titrated potentiometrically in the
presence of sodium and lithium in an acetic acid-chloroform medium
with standard perchloric acid dissolved in dioxane. Two inflections
are observed; the first one corresponds to potassium. Before titrating, the alkali halides, including fluorides, and the alkali sulfates
must be converted to the respective acetates. The method is not suitable for the determination of large amounts of potassium because of
the inconveniently large volumes necessary to keep the acetates in
solution. Nevertheless, the method affords a rapid and reasonably
accurate means for the analysis of binary mixtures of potassium with
sodium or lithium provided that the sum of the alkali metals is determined by standard methods, e.g., as sulfates or chlorides [3].

Method 2. An indirect, but very precise method, makes use of

the coulometric titration of tetraphenylborate ion with electrogenerated silver ion in aqueous acetone media. Small quantities (0.3-10.0
mg) of potassium are determined after precipitation as the tetraphenylborate. It is imperative that the washed precipitate be dissolved in
30-50 vol % of acetone while still wet since, upon drying, the tetraphenylborate undergoes hydrolytic decomposition which decreases the
tetraphenylborate of the precipitate by several percent. The precipitation is performed in a buffered acetate solution at a pH of 4.2
using sodium tetraphenylborate. The systematic error in the coulometric titration is not greater than about 0.4% [4].

Method 3. Potassium ion is determined by a potentiometric precipitation titration with tetraphenylborate ion using a cationic sensitive glass electrode as the indicator electrode. For this purpose,
a Beckman Zeromatic pH meter equipped with a 39137 cationic sensitive
glass electrode with a fiber junction calomel electrode is well suited.
Since the cationic sensitive electrode also responds to the sodium ion,
sodium tetraphenylboron must be converted to the calcium tetraphenyl
boron by ion exchange methods. This procedure provides a rapid simple
method for the direct determination of potassium. It has the advantage over other techniques of requiring relatively little equipment
and time, but it suffers from the disadvantage of being nonselective
with respect to individual ions in the alkali metal group [5].

Method 4. Potassium ion is titrated potentiometrically with a
standard solution of zinc hexafluosilicate (dissolved in water or
aqueous methanol) in a solution composed of 20% water and 80% methanol.
The end point is detected with a glass electrode having a composition
of 27% sodium oxide, 8% aluminum oxide, and 65% silicon dioxide. Calcium, magnesium, aluminum, and other ions precipitated by the hexafluosilicate, e.g., ammonium, sodium, and barium ions interfere [6].

References

[1]. R. E. Jentoff and R. J. Robinson, *Anal. Chem.*, *28*(12), 2011-2015
      (1956).

[2]. F. S. Sadek and C. N. Reilly, *Anal. Chem.*, *31*, (4), 494-498
      (1959).

[3]. L. I. Skresta and M. N. Das, *Anal. Chem.*, *39*, (11), 1300-1301
      (1957).

[4]. G. J. Patriarche and J. J. Lingane, *Anal. Chem.*, *39*(2), 168-171
      (1967).

[5]. G. A. Rechnitz, S. A. Katz and S. B. Zamochnick, *Anal. Chem.*,
      35(9), 1322-1323 (1963).

[6]. R. Geyer, I. Chonajnacki, W. Erxleben, and W. Syring, *Z. Anal.
      Chem.*, *204*(5), 325-331 (1964).

RUBIDIUM

The chemical behavior of rubidium is similar to that of cesium and potassium. There are no important classic methods to be reported for the titrimetric determination of rubidium. Modern techniques involve precipitation titrations.

Synoptic Survey

Classical Methods

None.

Contemporary Methods

Precipitation Titrations. Potentiometric titration of rubidium may be carried out using a solution of calcium tetraphenylboron [1].
Rubidium tetraphenylboron is titrated coulometrically with electrogenerated silver ion in an aqueous acetone media [2].

Outline of Recommended Contemporary Methods for Rubidium

Precipitation Titrations

Method 1. Rubidium is titrated potentiometrically with a standard solution of calcium tetraphenylboron using a fiber junction calomel electrode. The use of calcium tetraphenylboron is necessary since the cationic sensitive electrode also responds to the sodium ion. The available sodium tetraphenylboron may be converted to the calcium salt by ion exchange methods. During the titration the pH is maintained at 7.8. The method is rapid and simple. It is designed for the direct determination of univalent cations which form insoluble compounds with the tetraphenylborate ion. Titration of mixtures of the alkali metals give titration curves showing only a single inflection corresponding to the total alkali metal content. The method is thus nonselective but nevertheless requires little time and only a moderate amount of equipment. For 20.15 mg rubidium the relative error is 0.23%; for 40.40 mg rubidium the relative error is 0.00%; and for 60.40 mg rubidium the relative error is 0.00% [1].

Method 2. An indirect but very precise method makes use of the coulometric titration of the tetraphenylborate ion with electrogenerated silver ion in an aqueous-acetone media. Small samples containing 0.5-10 mg of rubidium are determined after precipitation as the tetraphenyl borate. It is imperative that the washed precipitate be dissolved in 30-50 vol % acetone while still wet. Drying of the rubidium tetraphenylboron results in a hydrolytic decomposition which decreases the tetraphenylborate content by several percent. The original precipitation is performed in a buffered acetate solution at pH 4.2 using sodium tetraphenylboron. The systematic error in the coulometric titration is not greater than about 0.4% [2].

References

[1]. G. A. Rechnitz, S. A. Katz, and S. B. Zamochnik, *Anal. Chem.*,
      35(9), 1322-1323 (1963).

[2]. G. J. Patriarche and J. J. Lingane, *Anal. Chem.*, *39*(2), 168-171
      (1967).

CESIUM

There are no important classic methods to be reported for the
titrimetric determination of cesium. The modern methodological approach is by redox, complexometric, or precipitation titration.

Synoptic Survey

Classical Methods

None of importance.

Contemporary Methods

Redox Titration. Cesium is precipitated as cesium permanganate.
The precipitate is dissolved in excess 0.1 N oxalic acid and the excess
acid is back-titrated with a standard permanganate solution [1].

Complexometric Titration. Cesium in the form of the chloride is
dissolved in water and the chloride is precipitated as silver chloride.
The latter is dissolved in ammoniacal potassium tetracyanonickel, and
the liberated nickel ion is titrated with standard EDTA [2].

Precipitation Titrations. Cesium is titrated potentiometrically
with standard calcium tetraphenylboron solution using a cation sensitive glass electrode [3].
Cesium is precipitated as cesium tetraphenylboron which is then
titrated coulometrically with electrogenerated silver ion in an aqueous acetone medium [4].
Cesium chloride dissolved in acetic acid is titrated conductometrically using antimony(III) chloride dissolved in acetic anhydride
as the titrant [5].

Outline of Recommended Contemporary Methods for Cesium

Redox Titration

Since cesium permanganate is sparingly soluble in water at low
temperatures (at 1.0° the solubility product is 1.5×10^{-5}), cesium
as cesium chloride is readily precipitated with ammonium permanganate.
The optimum sample contains 5-50 mg cesium. The precipitate is

dissolved in excess 0.1 N oxalic acid acidified with sulfuric acid.
The excess unreacted oxalic acid is back-titrated with standard potassium permanganate solution. Satisfactory results are obtained provided
that the concentrations of both reagent and cesium are sufficiently
high. However, the method is applicable only when the ratio of cesium
to rubidium is high. Potassium interferes to a much smaller extent.
Other metal ions except for silver do not interfere unless they are
able to reduce permanganate in a neutral solution [1].

Complexometric Titration

Cesium is precipitated as cesium perchlorate which in turn is
transformed into cesium chloride by heating with ammonium chloride.
The cesium chloride is subsequently dissolved in water, the chloride
precipitated as silver chloride, and, finally, the so-formed silver
chloride is dissolved in ammoniacal potassium tetracyanonickel.

$$2 \; AgCl \; + \; [Ni(CN)_4]^{-2} \; \rightarrow \; 2 \; [Ag(CN)_2]^{-1} \; + \; Ni^{+2} \; + \; 2 \; Cl^{-1}$$

The nickel ion which is formed according to the above equation is then
titrated with a 0.1 M EDTA solution using murexide as the indicator.
The percent error is 2.0% for 5.0 mg, 0.3% for 30 mg, and 0.5% for 100
mg cesium oxide [2].

Precipitation Titrations

Method 1. Cesium is titrated potentiometrically with a standard
calcium tetraphenylboron solution using a cation sensitive glass electrode opposite a fiber junction calomel electrode. Calcium tetraphenylboron is used as the titrant since the cation sensitive electrode
also responds to the sodium ion. Hence, a prior conversion of the more
readily available sodium tetraphenylboron to the calcium salt is necessary. The pH during the titration should be maintained at 7.8. The
method is both rapid and simple, and may be used for a variety of univalent cations which form insoluble compounds with the tetraphenylboron ion. The method requires relatively little equipment and time,
but the main disadvantage is its nonselectivity with regard to the ions
of the alkali metals. For 39.50 mg of cesium chloride the relative
error is 0.00%; for 79 mg of cesium chloride the relative error is
0.11%; and for 118.7 mg of cesium chloride the relative error is 0.025%
[3].

Method 2. An indirect but very precise method for the determination of cesium involves the precipitation of cesium tetraphenylboron followed by the titration of the tetraphenylboron ion with
electrogenerated silver ion in an aqueous acetone medium. Small samples containing 0.5-10.0 mg cesium are treated with sodium tetraphenylboron to precipitate the cesium as cesium tetraphenylboron.
This precipitate is then dissolved, while still wet, in 30-50 vol %
aqueous acetone. The precipitate after filtration should not be
allowed to become dry before dissolving it in the aqueous acetone since
the cesium tetraphenylboron readily undergoes hydrolytic decomposition.
The precipitation is carried out in an acetate buffered solution at
pH 4.2, and, after the precipitate of cesium tetraphenylboron is

dissolved in the aqueous acetone, the tetraphenylboron ion is titrated
with coulometrically generated silver ion. The systematic error in
this coulometric titration is not greater than 0.4% [4].

Method 3. Cesium is determined by the conductometric titration
of cesium chloride dissolved in acetic acid using as the titrant a
standard solution of antimony(III) chloride in acetic anhydride.

$$3 \; CsCl \; + \; 2 \; SbCl_3 \; \rightarrow \; Cs_3Sb_2Cl_9$$

This method is based upon the titration of a solution showing good
conductivity with a solution having poor conductivity. The optimum
concentration of cesium chloride is 6 mg/ml. Lithium and potassium do
not interfere. Iron(III) and rubidium ions must be absent. The relative error is as follows:

Wt. CsCl	Relative error (%)
315 mg	0.8
157 mg	0.4

Determinations carried out in the presence of other alkali metal halides (MX) show the following relative error [5]:

Wt. CsCl	Wt. MX present	Relative error (%)
315 mg	135 mg KCl	1.8
290 mg	23 mg NaCl	1.7

References

[1]. E. R. Caley and W. H. Deebel, *Anal. Chem.* *33*(2), 309-310 (1961).

[2]. A. de Sousa, *Talanta, 8,* 686-688 (1961).

[3]. G. A. Rechnitz, S. A. Katz, and S. B. Zamochnik, *Anal. Chem., 35*
(9), 1322-1323 (1963).

[4]. G. J. Patriarche and J. J. Lingane, *Anal. Chem., 39*(2), 168-171
(1967).

[5]. J. Harvir, *Collection of Czech. Chem. Commun., 25,* 695-700
(1960).

COPPER

The preferred current methods for the determination of copper
include redox, complexometric, and precipitation titration.

Synoptic Survey

Classical Methods

Redox Titration. Copper(I) may be titrated with standard cerate
solutions [1], amperometrically with standard potassium dichromate
solutions [2], or after precipitation as the thiocyanate, the titration
may be carried out using potassium iodate solutions [3]. Copper(II)
may be determined iodometrically by titrating the iodine liberated from
solutions of KI with $Na_2S_2O_3$ [4].

Contemporary Methods

Redox Titrations. The classic iodometric copper determination
with sodium thiosulfate is carried out as a thermometric titration
[5].
The reduction of copper(II) to copper(I) is utilized in the titration with electrogenerated tin(II) [6].
Copper(II) may be determined by conductometric titration with
sulfhydryl groups which are generated internally from the stable mercury complex of thioglycolic acid [7].
Copper(I), obtained from copper(II) by reduction with disodium
phosphite in acidic solutions, is titrated with standard potassium
ferricyanide [8].
A potentiometric titration based on a stepwise reduction of ti-
tanium(III) chloride is used to determine copper(I), copper(II), iron
(II), and iron(III) without intermediate separation [9].

Complexometric Titrations. Copper(II) is titrated potentiometrically with tetraethylenepentamine (Tetren). This method is
superior to the titration using Trien (triethylenetetramine) [10].
In a mixture of silver and copper(II) the copper is titrated
with the barium complex of EDTA [11].

Precipitation Titration. Copper(II) is reduced to copper(I) with
ascorbic acid and then titrated amperometrically with tetraphenylboron
[12].

Outline of Recommended Contemporary Methods for Copper

Redox Titrations

Method 1. Copper(II) is determined iodometrically by a thermometric titration. The titration of iodine liberated by the reaction
of excess iodide ions and cupric ions with a standard thiosulfate
solution causes a temperature rise in the reacting mixture. A

rectilinear ascending type curve is observed. The equivalence point
is characterized by a sharp break in the curve followed by a straight
negatively sloped excess reagent line. This method has the advantage
of being rapid and easily convertible for automatic titration. The
precision and accuracy vary between 1 and 3% [5].

Method 2. Copper(II) ion is titrated to copper(I) with electrogenerated tin(II) ion. The supporting electrolyte consists of 4 M
sodium bromide and 0.2 M ammonium chloride. A platinum generator
electrode is used. The end point is determined potentiometrically.
Quantities of copper from 1-35 mg in a volume of 90 ml can be determined with an error of $\pm$ 0.3%. Many of the elements commonly assoc-
iated with copper do not interfere [6].

Method 3. The sulfhydryl group has a great affinity for metals.
It is also an active reducing agent and is of wide applicability in
quantitative analysis. Many metals which cannot be determined by EDTA
can be profitably determined using a reagent carrying a sulfhydryl
grouping. This includes such elements as the platinum metals, silver,
and gold. This reagent may also be used in the determination of the
thioanions-forming elements of the acid hydrogen sulfide group. As a
reductant, the sulfhydryl group has the advantage of being a storage-
reducing agent in the high pH range where there are few good reducing
agents available. Since organically bound sulfhydryl groups are unstable towards air oxidation, and since they have an objectionable
odor, internal generation of them by constant current titrimetry is
the best approach. The very stable mercury complex of thioglycolic
acid is probably the best source for the sulfhydryl group. The response of a mercury pM electrode or amperometry at two mercury electrodes is used for detection of the end point. The reaction probably
proceeds according to the equation:

$$2 \; Cu(II) \;+\; 2 \; RSH \;\rightarrow\; 2 \; Cu(I) \;+\; RSSR \;+\; 2 \; H^{+}$$

Copper can be determined directly in brass and chrome alloys without
prior separation. For such alloys where the copper content averages
73-88%, the standard deviation is 0.4%, and the error is $\pm$ 0.56% [7].

Method 4. Copper(II) is determined by reduction to copper(I)
using sodium hypophosphite and subsequently titrated with ferricyanide.
The reduction of the copper(II) in samples containing 0.1-1.5 mmoles
of copper proceeds at room temperature and reaches completion in 1-2
minutes. The reduction is carried out in the presence of excess
hydrochloric acid ($\geq$ 4 N), and for the above sized sample, 0.5-1 gm
of sodium hypo phosphite monohydrate is used.

$$2 \; HCuCl_3 \;+\; NaH_2PO_2 \;+\; H_2O \;\rightarrow\; 2 \; HCuCl_2 \;+\; NaH_2PO_3 \;+\; 2 \; HCl$$

The cuprous ion which is formed is titrated potentiometrically with
0.1 N potassium ferricyanide.

$$3 \; HCuCl_2 \;+\; 3 \; K_3[Fe(CN)_6] \;\rightarrow\; K_2Cu_3[Fe(CN)_6]_2 \;+\; K_4[Fe(CN)]_6$$
$$+\; 3 \; HCl \;+\; 3 \; KCl$$

Excess hypophosphite present in the solution cannot reduce the copper (II) in the potassium-copper ferrocyanide. The titration of copper(I) with the usual oxidants such as potassium dichromate or vanadic acid gives high or irreproducible results. The titration is performed in air since the excess hypophosphite in the acidic solution prevents the air or dissolved oxygen oxidation of copper(I) to copper(II). The accuracy of the method is $\pm$ 1%. In the presence of nickel, cobalt, iron, or zinc the results are high because of the possible formation of insoluble ferricyanides. The method may be used for the determination of copper in plating baths since the products of oxidation or decomposition of formaldehyde such as the carboxyl group and methyl alcohol which are usually present in the baths do not interfere [8].

Method 5. Copper(I), copper(II), iron(II), and iron(III) present as their respective chlorocomplexes in solutions containing 2 M chloride ion may be determined by potentiometric titration using titanium (III) chloride as the titrant. In one portion of the solution, the chloride ion concentration is adjusted to 2.5 M, and the iron(III) is titrated with 0.05 N titanium(III) chloride. After the addition of 5 ml of 10% ammonium thiocyanate, the copper(II) is titrated in the same sample. The sum of the iron and copper is determined in a second portion following oxidation with 0.05 M potassium dichromate solution. The difference corresponds to iron(II) and copper(I), respectively. The method is applicable to the determination of small amounts of iron in the presence of a 25-fold amount of copper [9].

Complexometric Titrations

Method 1. Based on theoretical and practical considerations, Tetren is superior to Trien as a selective titrant for metal ions such as copper(II), mercury(II), nickel(II), zinc(II), and cadmium(II). In the determination of copper(II), it is titrated potentiometrically alone or in the presence of nickel, zinc, and cadmium at low pH using a mercury electrode for end-point detection. The rare earth elements, the alkaline earth elements, aluminum, bismuth, lead, and scandium do not interfere. Standard deviations are as follows [10]:

Wt. copper	Wt. other metal present	Deviation
7.62 mg	6.86 mg zinc	0.1
7.92 mg	7.74 mg nickel	0.25
6.36 mg	12.48 mg cadmium	0.4

Method 2. Copper(II) may be determined in a mixture of silver and copper(II) ions either before or after the precipitation of the silver as the bromide. If the silver ion is titrated first, then care must be taken to add only a 1-2 drop excess of 0.1 M bromide ion. The copper ions in the solution are titrated with Na_2BaL (H_4L = EDTA) with 2-3 drops of HgL^{-2} as the indicator. A platinum indicator electrode system responsive to copper and mercury potentials is used to detect the end point. As a chelating agent, Na_2BaL has

the advantage that no buffer is required since during the titration no protons are liberated [11].

Precipitation Titration

Copper(I) is titrated amperometrically using a dropping mercury electrode. The copper concentration in the solution is 0.01 M and the titrant is a 0.01 M solution of tetraphenylboron ion. Before titration, the copper(II) ions are reduced to copper(I) by the addition of solid ascorbic acid. Lead, arsenic, antimony, tin, nickel, and cobalt do not interfere, but silver and thallium, as well as sodium and lithium do interfere in high concentrations. Precision of the method is ± 0.3%. Due to the specificity of the tetraphenylboron ion, this method is probably the best and simplest method for the determination of copper in most common mixtures [12].

References

[1]. N. H. Furman, (ed.), *Standard Methods of Chemical Analysis*, 6th Ed., D. Van Nostrand, Princeton, N. J., 1962, Vol. 1, pp. 405-406.

[2]. N. H. Furman, (ed.), *Standard Methods of Chemical Analysis*, 6th Ed., D. Van Nostrand, Princeton, N. J., 1962, Vol. 1, p. 406.

[3]. N. H. Furman, (ed.), *Standard Methods of Chemical Analysis*, 6th Ed., D. Van Nostrand, Princeton, N. J., 1962, Vol. 1, p. 405.

[4]. N. H. Furman, (ed.), *Standard Methods of Chemical Analysis*, 6th Ed., D. Van Nostrand, Princeton, N. J., 1962, Vol. 1, p. 404.

[5]. E. J. Billingham, Jr. and A. H. Reed, *Anal. Chem.*, *36*(6), 1148-1149 (1964).

[6]. J. J. Lingane, *Anal. Chim. Acta.*, *21*, 227-232 (1959).

[7]. B. Miller and D. N. Hume, *Anal. Chem.*, *32*(4), 524-530 (1960).

[8]. P. Norkus and R. Markoviciene, *Lietuvos TSR Mokslu Akad. Darbai Ser. B.*, *1967*(2), 89-95.

[9]. L. Sucha and Z. Urner, *Collection Czech. Commun.*, *29*(7), 1612-1617 (1964).

[10]. C. N. Reilly and A. Vavonlis, *Anal. Chem.*, *31*, 243-248 (1959).

[11]. F. Sierra and J. Hernandez-Canavate, *Anal. Real Soc. Espan. Fis. Quim. (Madrid) Ser. B.*, *63*(7-8), 821-826 (1967).

[12]. A. M. Hartley and A. Hehner, *Collection Czech. Chem. Commun.*, *30*, 121, 4250-4256 (1965).

SILVER

Current methods for the determination of silver involve acid-base, complexometric, and precipitation titrations.

Synoptic Survey

Classical Methods

Complexometric Titration. Silver is precipitated as the chloride, dissolved in excess KCN, and titrated with silver nitrate [1].

Precipitation Titration. Silver is titrated with standard thiocyanate [2], or with standard bromide solution [3].

Contemporary Methods

Acid-Base Titrations. Small amounts of silver are titrated with standard thioglycolic acid [4]. Silver in samples containing a few parts per million is titrated potentiometrically with standard dithiooxamide [5].

Complexometric Titrations. Silver ion is reduced with metallic copper. The equivalent amount of copper(II) ion formed is titrated with standard EDTA [6].
Silver is titrated coulometrically. The mercury(II) complex of monothioethylene glycol (MTEG) is used as the source of the complexing sulfhydryl groups [7].
Silver is titrated amperometrically with Unithiol (sodium 2,3-dimercapto-1-propane sulfonate). The latter forms complexes with both silver and palladium(II) [8].

Precipitation Titration. Very small amounts of silver usually in the range of parts per million are determined by titration with standard p-dimethylaminobenzylidenerhodamine. The titration is carried out in a CCl_4 solution. A slight excess of titrant colors the organic phase yellow-orange [9].

Outline of Recommended Contemporary Methods for Silver

Acid-Base Titrations

Method 1. Silver samples containing 10^{-1} to $10^{-4}\gamma$ are titrated potentiometrically with standard thioglycolic acid. Two breaks in the titration curve are observed corresponding to the formation of $AgSCH_2COOAg$ and $(AgSCH_2COOH)_2$. If the first break is taken as the end point, then copper(II) does not interfere and the method can, therefore, be readily used for the determination of silver in silver-copper alloys [4].

Method 2. Small amounts of silver (0.07-50 ppm) are titrated potentiometrically with a standard aqueous solution of dithiooxamide using a pH meter equipped with a silver indicator and a glass reference electrode. Copper, zinc, cadmium, nickel, cobalt, or lead do not interfere. Due to the greater potential break with dithiooxamide, the results are more precise for very low concentration of silver. The method has the following advantages: (1) few interferences, (2) no preliminary separations, (3) rapidity, (4) high sensitivity, (5) less subject to variables which affect the precision and accuracy. The average accuracy is about 0.5% and the relative standard deviation ranges from 0.03 to 1.26% [5].

Complexometric Titrations

Method 1. Silver samples containing 0.2-1 mmole are determined indirectly by reducing silver ions with metallic copper. The copper (II) ions so formed are titrated with standard EDTA solution using murexide as the indicator. The deviation varies between 0.16% for 29.58 mg of silver to 0.06% for 107.9 mg of silver. In the latter case, the determination was carried out in the presence of 4 mmoles of potassium nitrate.

Method 2. Silver is titrated with MTEG. The latter has sufficient water solubility to permit a constant current coulometric generation of sulfhydryl groups. Interferences from secondary complexes involving the solubilizing hydroxyl group are minimum for this compound. These features permit the use of sulfhydryl-titration in the determination of silver, palladium(II), and platinum(II). Precision and accuracy are excellent [7].

Method 3. Silver and palladium form with Unithiol a complex of the composition $PdAg_2(Un)_2^{-2}$ where Un is $C_3H_5O_3S_3^{-3}$. Silver and palladium may be titrated amperometrically, either separately or together. The titration is also feasible in the presence of 1000-fold amounts of zinc, cadmium, lead, or bismuth. Gold, iron, and copper interfere. The interferences from iron(III) may be eliminated by complexing it with pyrophosphate or ammonium fluoride [8].

Precipitation Titrations

Method 1. Very small amounts of silver are determined by three-phase titrimetry involving an organic, a solid, and an aqueous phase. The procedure is based on the fact that the precipitate formed upon titrating the aqueous silver sample with p-dimethylaminobenzylidenerhodamine in CCl_4 is separated rapidly and quantitatively at the boundary of the two liquid phases. Upon shaking the mixture, the aqueous layer becomes colorless within a very short time, and the colorless organic phase will show an orange-yellow color only with the addition of a very slight excess of the titrant. Silver samples containing 2-20 μ may be determined. An individual titration requires 10-20 minutes. The relative error varies between ± 0.4 to ± 0.8% [9].

References

[1]. N. H. Furman, ed., *Standard Methods of Chemical Analysis*, 6th Ed., D. Van Nostrand, Princeton, N. J., 1962, Vol. 1, p. 983.

[2]. N. H. Furman, ed., *Standard Methods of Chemical Analysis*, 6th Ed., D. Van Nostrand, Princeton, N. J., 1962, Vol. 1, p. 983.

[3]. K. Fajans and H. Wolff, *Z. Anorg. Allg. Chem. 137*, 241 (1924).

[4]. Y. Takeuchi, *Nippon Kagaku Zasshi, 72*, 1644-1677 (1961).

[5]. L. H. Kalbus and G. E. Kalbus, *Anal. Chim. Acta, 39*, 335-340 (1967).

[6]. C. Heunart, *Talanta, 12*(17), 694-695 (1965).

[7]. B. Miller and D. N. Hume, *Anal. Chem., 32*(7), 764-767 (1960).

[8]. D. A. Songina, Kh. K. Ospanov, Z. B. Rozhdestrenskaya, and T. K. Gutermakher, *Zh. Analit. Khim.*, 22(8), 1170-1174 (1967).

[9]. H. Lux, T. Niedermaier, and K. Petz, *Z. Anal. Chem., 171*, 173-176 (1959).

GOLD

Contemporary methods for the determination of gold include redox and complexometric titrations.

Synoptic Survey

Classical Methods

Redox Titration. Gold may be determined iodometrically in which the iodine liberated from KI by gold(III) ion is titrated with standard arsenite solutions [1,2]. Gold may also be determined by titration potentiometrically with tin(II) chloride [3] or with copper(I) chloride [4].

Contemporary Methods

Redox Titrations. Gold(III) is titrated with standard ascorbic acid in hydrochloric acid. Variamine blue is used as the indicator [5].

Gold(III) is reduced with excess standard manganese(II) in a pyrophosphate solution. The excess is then back-titrated with standard permanganate [6]. Gold is titrated coulometrically with electrogenerated tin(II) [7,8].

Complexometric Titrations. Gold is titrated with standard Unithiol (sodium 2,3-dimercapto-1-propane sulfonate). The minimum gold
concentration is 50 γ/25 ml [9].

Gold(III) is titrated coulometrically with electrogenerated sulfhydryl groups derived from thioglycolic acid [10] or from monothioethylene glycol (MTEG) [11].

Outline of Recommended Contemporary Methods for Gold

Redox Titrations

Method 1. Gold(III) is titrated with 0.01 or 0.001 M ascorbic
acid solution in hydrochloric acid in the presence of potassium bromide. The chloride ion concentration should be less than one molar
and the optimum temperature for the titration is 50°. The end point
is recognized by the disappearance of the orange color of the tetra-
bromoauric ion. Approximately 500γ of gold can be extracted with
ethyl acetate and titrated with standard ascorbic acid in the presence
of platinum, copper, selenium, tellurium, lead, etc. Variamine blue
is used as the indicator. The error varies between 2 to 5% [5].

Method 2. Gold(III) is reduced by manganous ions in a pyrophosphate solution according to the equation:

$$Au^{+3} + 3\ Mn^{+2} + 9\ H_2P_2O_7^{-2} \rightarrow Au + 3[Mn(H_2P_2O_7)_3]^{-3}$$

In using this method for the determination of gold, the excess of the
standard manganous solution is back-titrated potentiometrically with
a standard potassium permanganate solution. A large excess of pyrophosphate is necessary in the presence of certain cations which form
precipitates. Iron(III) is masked by the addition of hydrofluoric
acid. Gold can be determined in the presence of tenfold amounts of
the following ions: nickel(II), cobalt(II), iron(III), lead(II),
iridium(III), rhodium(III), chloride, sulfate, nitrate, phosphate,
and perchlorate. Reducing agents such as iodide, bromide, and hexa-
chloroplatinate ions interfere [6].

Method 3. Gold(III) is titrated with electrogenerated tin(II)
in a supporting electrolyte consisting of 3-4 M NaBr, 0.3 N HCl, and
0.2 M tin(IV) chloride. The end point is detected potentiometrically,
amperometrically, or spectrophotometrically at a wavelength of 400 mμ.
In either a potentiometric or an amperometric titration, 0.5-23.0 mg
of gold can be determined with an average error of about $\pm$ 0.3%. The
average error in the spectrophotometric titration is about $\pm$ 2%.

Complexometric Titrations

Method 1. Gold is titrated with a standard Unithiol solution
in 0.5-4 N hydrochloric acid. The presence of nitric acid in amounts
up to 3 vol % or of lead, zinc, silver, or bismuth does not interfere.
If iron is present in quantities equal to or larger than the gold,
ammonium fluoride must be added. The presence of copper ion in 100-
fold excess improves the determination. In higher concentrations,

the results are low. The minimum gold concentration which should be determined by this method is 50 γ in 25 ml of solution [9].

Method 2. Gold(III) is titrated coulometrically with an electro-generated sulfhydryl group derived from $HSCH_2COOH$. The reagent used is the very stable mercuric complex of thioglycolic acid which is soluble at pH values above the pK_1 of the acid, i.e. 3.60. Free sulfhydryl groups are produced according to the equation:

$$Hg(SR)_2 \ + \ 2 \ H^+ \ + \ 2e^- \ \rightarrow \ Hg \ + \ 2 \ RSH$$

$$(R \ = \ -CH_2COOH)$$

A mercury pM electrode is used for end point detection. Amperometry at two mercury electrodes is also feasible. The error is $\pm$ 0.3% for 2.011 mg of gold [10].

Method 3. Gold(III) is titrated coulometrically with an electro-generated sulfhydryl group from monothioethylene glycol (MTEG), $HSCH_2$-CH_2OH. This compound is sufficiently soluble in water to permit constant current coulometric generation of its sulfhydryl group. Interferences from secondary complexes involving the solubilizing hydroxyl group are minimal for this compound; MTEG has been recommended as being preferable to thioglycolic acid as a source for sulfhydryl groups. An accuracy of $\pm$ 0.5% is easily achieved in the determination of gold in samples containing 200 γ to 2 mg of gold [11].

References

[1]. W. B. Pollard, *Bull. Inst. Mining and Met. No. 330-331*, 23 (1932).

[2]. V. E. Heschlag, *Ind. Eng. Chem., Anal. Ed., 13*, 561(1941).

[3]. E. Mueller and R. Bensewitz, *Z. Anorg. Allgem. Chem. 179*, 113 (1929).

[4]. E. Mueller and K. H. Ranzler, *Z. Anal. Chem., 89*, 229(1967).

[5]. G. Rady and L. Erdey, *Talanta, 14*(3), 425-429 (1967).

[6]. M. Kotoucek, J. Dolezal, and J. Zyka, *Collection Czech. Chem. Commun., 28*, 521-524 (1963).

[7]. A. J. Bard and J. J. Lingane, *Anal. Chim. Acta, 20*, 581-587 (1959).

[8]. A. J. Bard and J. J. Lingane, *Anal. Chim. Acta, 20*, 463 (1959).

[9]. O. A. Songina, Kh. K. Ospanov, and Z. B. Rozhdestrenskaya, *Zavodsk. Lab., 30*(6), 664-667 (1964).

[10]. B. Miller and D. N. Hume, *Anal. Chem.*, *32*(4), 524–528 (1960).

[11]. B. Miller and D. N. Hume, *Anal. Chem.*, *32*(7), 564–526 (1960).

GROUP II

Beryllium	Barium
Magnesium	Zinc
Calcium	Cadmium
Strontium	Mercury

BERYLLIUM

Methods for the determination of beryllium include acid-base and complexometric titration.

Synoptic Survey

Classical Methods

Acid-Base Titrations. The hydroxyl ions liberated by the addition of excess sodium or potassium fluoride to beryllium hydroxide are titrated with a standard acid [1].

Beryllium is titrated with sodium hydroxide. A high frequency titrimeter serves to indicate the end point [2].

Contemporary Methods

Acid-Base Titration. The classic methods above have been modified to permit the determination of beryllium in the presence of uranium. Complexation of the uranium is carried out using hydrogen peroxide to form peroxyuranium compounds, followed by titration with standard acid [3].

Complexometric Titrations. Excess sodium sulfosalicylate is added to the sample solution containing the beryllium. This excess is back-titrated with standard beryllium sulfate [4].

The complex cobalt-beryllium compound formed by the reaction of hexamminecobaltic trichloride solution with beryllium carbonate is decomposed, and its cobalt content is determined by EDTA titration [5].

Hexamminecobalt(III) hexacarbonato-oxo-tetraberyllate is decomposed by fuming with sulfuric acid. The cobalt(II) formed is determined by adding excess standard EDTA and back-titrating this excess with 0.01 M cobalt(II) [6].

Outline of Recommended Contemporary Methods for Beryllium

Acid-Base Titration

    The classic titrimetric method for the determination of beryllium involves the addition of fluoride ion and the titration of the liberated hydroxyl ions with standard acid.

$$Be(OH)_2 + 4 F^- \rightarrow (BeF_4)^{-2} + 2 OH^-$$

This method cannot be used in the presence of uranium ions because of the simultaneous precipitation of both the beryllium and the uranium hydroxides. To remove this interference it is necessary to form a complex of uranium with a larger pK value than that of the complex formed with the fluoride ion. The titration can thus be carried out by the addition of hydrogen peroxide to the solution forming peroxyuranium compounds which do not interfere with the titration of the beryllium.

$$H_2UO_4 + 3 H_2O_2 \rightarrow "H_4UO_8" + 2 H_2O$$

Values obtained in the titration of beryllium in the presence of uranium are very satisfactory. The precision is 1-3% for 2 to 20 mg of beryllium [3].

Complexometric Titrations

    Method 1. Sulfosalicyclic acid forms a well defined molecular species with beryllium. The complex is very stable.

$$Be(sulfosal.)_2^{-4} \quad K = 2.2 \times 10^{20}$$

The reaction between beryllium and sulfosalicylate is stoichiometric (1 to 2). To the sample a known excess of sodium sulfosalicylate is added, and this excess is then back-titrated with standard beryllium sulfate solution using a photometric titrator and the trisodium salt of 3-(2-arsonophenylazo)-4,5-dihydroxy-2, 7-naphthalene disulfonic acid as the indicator. The end point occurs at a pH of 10.6. The interference of most elements is prevented by the addition of EDTA although at times a better masking can be obtained using CyDTA (1,2-cyclohexylenedinitrilo-tetraacetic acid). Uranium is masked with hydrogen peroxide; chromium is oxidized to the hexavalent state. The titration range is from 0.05 to 15 mg of beryllium with a relative standard deviation of $\pm$ 0.13% at the 3 mg beryllium level, and $\pm$ 0.38% for samples containing 0.8 mg of beryllium [4].

    Method 2. The addition of hexamminecobalt(III) chloride solution to a solution of beryllium in excess ammonium carbonate causes the precipitation of the complex $[Co(NH_3)](H_2O)_2 Be_2(CO_3)_2(OH)_3]$. $3H_2O$. Samples containing beryllium are thus precipitated, and the precipitate containing the beryllium is removed by filtration, washed, and decomposed with sulfuric acid. The cobalt so liberated is titrated with standard EDTA solution with murexide indicator. Using EDTA as a masking agent, beryllium can be separated and determined in the presence of iron(III), aluminum and magnesium [5].

Method 3. A third but excellent complexometric method is based on the precipitation of hexamminecobalt(III) hexacarbonate-oxo-tetra-beryllate.

$$[Co(NH_3)_6]_2 \ [Be_4O(CO_3)_6].xH_2O$$

This precipitate is decomposed by fuming with sulfuric acid, and the cobalt(II) liberated is determined by weight titration with 0.01 M EDTA and back-titration with 0.01 M cobalt(II) solution. The titration is performed at a pH of 8.5 to 9.0 in a potassium carbonate/hydrogen carbonate buffer in the presence of potassium thiocyanate, tetraphenylarsonium chloride, and chloroform. The end point is indicated by the appearance in the chloroform layer of the blue color of the ion association pair tetraphenylarsonium-tetrathiocyanato-cobaltate(II). The precision is 0.02% [6].

References

[1]. N. H. Furman, *Standard Methods of Chemical Analysis*, 6th Ed., Vol. 1, D. Van Nostrand, Princeton, N. J., 1962, pp. 168-171.

[2]. K. Anderson and D. Revinson, *Anal. Chem.*, *33*(10), 1272-1274 (1950).

[3]. J. V. Dagh and J. Spitz, *Anal. Chim. Acta*, *41*, 173-175 (1968).

[4]. T. M. Florence and Y. J. Farrar, *Anal. Chem.*, *35*(6), 712-714 (1963).

[5]. S. Misumi and T. Taketatsu, *Bull. Chem. Soc. Japan*, *32*, 593-596 (1959).

[6]. R. G. Monk and K. A. Exelby, *Talanta*, *12*(1), 91-100 (1965).

MAGNESIUM

Methods for the determination of magnesium include acid-base and complexometric titrations.

Synoptic Survey

Classical Methods

Acid-Base Titration. Magnesium is precipitated as magnesium ammonium phosphate. The precipitate is dissolved in excess sulfuric acid, and the excess acid is back-titrated with standard base [1].

Complexometric Titration. Magnesium is titrated with standard
EDTA. The end point is determined spectrophotometrically [2].

Contemporary Methods

Complexometric Titrations. Magnesium is titrated with 0.01 M
diethylenetrinitrilopentaacetic acid (DTPA) to an Eriochrome Black T
end point [3].

Magnesium is titrated with standard 1,2-cyclohexanedinitrilo-
tetraacetic acid (CDTA) using methyl-thymol blue as the indicator [4].

Magnesium is titrated in the presence of calcium by first masking
the calcium with ethylene-glycolbis(β-aminoethylether)-N,N-tetraacetic
acid (EGTA). The magnesium is then titrated as usual with EDTA [5].

Magnesium down to 0.03% magnesium oxide in the presence of as
high as 70% phosphorus pentoxide is titrated with CDTA [6,7].

Outline of Recommended Contemporary Methods for Magnesium

Complexometric Titrations

Method 1. The accuracy of the complexometric titration of magnesium is improved by using DTPA in place of EDTA. Titrations are
carried out using 0.01 M DTPA with Eriochrome Black T as the indicator.
In contrast to the titration of barium, it is unimportant whether Mg-
DTPA is present or not. In either case, the titration is accurate to
0.02-0.3% with relative standard deviations of 0.3% or less [3].

Method 2. Magnesium is titrated with standard CDTA using methylthymol blue as the indicator. Calcium is masked with excess standard
EGTA. Calcium may then be determined in the same solution by back-
titration of the excess EGTA with standard calcium chloride solution.
Compared with other procedures, this method has the advantage of per-
mitting the use of a variety of masking reagents (triethanolamine,
potassium cyanide, thioglycolic acid, etc.) for masking other elements
[4].

Method 3. Due to differences in the stability constants of the
magnesium, calcium, and barium complexes of EGTA, it is possible to
mask calcium and then titrate magnesium with standard EDTA. These
stability constants are as follows:

$$Mg-EGTA \qquad \log K = 5.21$$

$$Ca-EGTA \qquad \log K = 10.97$$

$$Ba-EGTA \qquad \log K = 8.41$$

Solutions containing magnesium and calcium are treated with Ba-EGTA.
This results in a displacement of the barium from the complex and the
formation of Ca-EGTA.

$$Mg^{+2} + Ca^{+2} + Ba-EGTA \rightarrow Ca-EGTA + Ba^{+2} + Mg^{+2}$$

The liberated barium ion is precipitated as barium sulfate, and the magnesium is then titrated with standard EDTA in an ammoniacal solution using an Eriochrome Black T indicator. By this procedure, magnesium can be determined in the presence of a very large excess of calcium. Furthermore, the separation of the barium sulfate precipitate is not only precise but rapid [5].

Method 4. Magnesium is determined accurately in phosphate rock and phosphoric acid down to 0.03% magnesium oxide in the presence of 0-70% phosphorus pentoxide by titration with standard CDTA. It is advantageous to substitute CDTA for EDTA because the pertinent equilibria of the metal-CDTA complexes are greater by the order of 10^2 compared to the respective metal-EDTA complexes. This increased stability results in consistently sharp complex formation and end-point detection. Interfering metals are first masked with CDTA in acid solution. The pH of the mixture is then adjusted to 5.0, and the excess CDTA is titrated with standard zinc solution using xylenol orange as the indicator. The total metal ion concentration is determined by titration in the same manner at pH 10. In this case the excess CDTA is back-titrated with standard magnesium solution using Eriochrome Black T as the indicator. Calcium is determined with a calcein II modified indicator under ultraviolet light by titration with CDTA at pH > 12. Magnesium is calculated with an unequivocal linear correction being made for interfering heavy metal ions. The determination may be carried out in less than 20 minutes after sample digestion for Ca:Mg ratios from 1:9 to 850:1 [6]. The method is greatly improved by employing spectrophotometric end-point detection [7].

References

[1]. J. O. Handy, *J. Am. Chem. Soc.*, *22*, 31 (1900).

[2]. P. B. Sweetser and E. D. Bricker, *Anal. Chem.*, *26*(1), 195 (1954).

[3]. E. D. Olson and R. J. Novak, *Anal. Chem.*, *38*(1), 152-153 (1966).

[4]. R. Pribil and V. Vesely, *Talanta*, *13*(2), 233-236 (1961).

[5]. R. Fabrega, A. Badrinas and A. Prieto, *Talanta*, *8*, 804-808 (1961).

[6]. D. C. Jordan and D. E. Monn, *Anal. Chim. Acta*, *37*, 42-48 (1967).

[7]. D. C. Jordan and D. E. Monn, *Anal. Chim. Acta*, *39*, 401-404 (1967).

CALCIUM

Methods for the determination of calcium include indirect redox, complexometric, and precipitation titrations.

Synoptic Survey

Classical Methods

Redox Titration. Calcium is separated as the sparingly soluble calcium oxalate monohydrate. The precipitated oxalate is then titrated with standard potassium permanganate [1].

Contemporary Methods

Complexometric Titrations. Calcium is determined, especially in water, by EDTA titration using 3'3" bis(2-hydroxy-3-carboxy-naphthalene-azophenophthalein as the indicator [2].

Calcium is titrated amperometrically with EDTA in the presence of magnesium at a pH 12-13 [3].

Small amounts of calcium are titrated biamperometrically with EDTA [4].

Calcium in the presence of magnesium is titrated amperometrically with ethyleneglycol bis(2-aminoethyl-ether)tetraacetic acid, EGTA [5].

Calcium in microgram quantities and in the presence of magnesium is titrated with coulometrically generated EGTA [6].

Precipitation Titration. Calcium in the presence of magnesium is determined by automatic thermometric titration with oxalate [7].

Outline of Recommended Contemporary Methods for Calcium

Complexometric Titrations

Method 1. In the successive titration of calcium and magnesium with EDTA, especially in water analysis, 3',3" bis(2-hydroxy-3-carboxy-naphthaleneazo)phenolphthalein (BNP) is used to advantage as the indicator. At the end point the color change is from blue to red-violet. The titration is carried out at pH 13. The method is far more convenient and faster than the ordinary chelation method especially for the determination of calcium and magnesium in water [2].

Method 2. Calcium is titrated amperometrically in the presence of magnesium with standard EDTA solution at pH 12-13. Using a potential of 0.7-0.73 V versus SCE on the platinum electrode, 0.0002 to 0.16 gm of calcium in 50 ml can be determined. Under these conditions, the amount of magnesium in the same solution can vary up to 40 times the calcium concentration. When the calcium concentration of the solution is 0.008 to 0.07 gm per 50 ml, then the concentration of magnesium must be ten times this amount. Otherwise, the results are low because of the layering of magnesium hydroxide. During the titration the concentration of ammonium ion, aluminum ion, and silver ion can be 400, 24, and 100 times as great as that of the calcium ion, respectively. The presence of sodium silicate improves the character of the curve. The amount of iron should be less than 40% of the calcium concentration. Calcite, blast-furnace slag, and open hearth slag are analyzed by this method. The error varies between 0.4 and 0.64% [3].

Method 3. Small amounts of calcium ($\geq$ 5 γ) are determined by biamperometric titration with standard EDTA solution. Magnesium does not interfere up to a magnesium-calcium ratio of 15:1. The titration is carried out automatically. The method is useful for the determination of very small amounts of calcium in water analysis. The percent differences found by visual and biamperometric titration of 0.011 mg of calcium in the presence of 1.000 to 20.00 mg of magnesium is $\pm$ 1.00 [4].

Method 4. In a mixture of calcium and magnesium, the calcium is first titrated amperometrically with standard EGTA followed by the titration of the magnesium with standard EDTA. It is possible to determine with good precision as little as 0.01 μg of magnesium in the presence of 0.01 mg of calcium per milliliter, and 0.01 μg of calcium in the presence of 0.01 mg of magnesium per milliliter. The method is applicable for the analysis of blood serum, urine, water, cement, etc. [5].

Method 5. Microgram quantities of calcium are titrated in the presence of magnesium with coulometrically generated EGTA. A potentiometric end point is used. As little as 40 μg of calcium can be titrated with a 4% average error and a standard deviation of $\pm$ 1.4 μg [6].

Precipitation Titration

Under controlled experimental conditions, calcium and magnesium in dilute buffered solutions yield strikingly different thermometric curves upon the addition of ammonium oxalate. Calcium is thus determined by automatic thermometric titration in the presence of magnesium by titration with standard oxalate solution. The method is ideally suited for rapid and convenient calcium determination in limestone and dolomite. Precision and accuracy is 0.5% in the concentration ranges between 5 x 10^{-3} M and 1 x 10^{-1} M [7].

References

[1]. N. H. Furman, ed., *Standard Methods of Chemical Analysis*, 6th Ed., D. Van Nostrand, Vol. 1, Princeton, N. J., 1962, p. 264.

[2]. K. Toei and T. Kobatake, *Talanta*, *14*(11), 1354 (1967).

[3]. M. A. Vitkina and G. E. Bekleshova, *Sovrem Metody Khim. Spektral. Anal. Mater.*, *1967*, 237-240.

[4]. J. Vorlicek, M. Fara, and F. Vydra, *Microchem J.*, *12*(3) 409-414 (1967).

[5]. D. Monnier and A. Roneche, *Helv. Chim. Acta*, *47*(1) 103-111 (1964).

[6]. G. D. Christian, E. C. Knoblock, and W. C. Purdy, *Anal. Chem.*, *37*(2), 292-294 (1965).

[7].   J. Jordan and E. J. Billingham, *Anal. Chem.*, *33*, 120-123 (1961).

STRONTIUM

Methods for the determination of strontium include acid-base, complexometric, and precipitation titrations. Classic methods depend on indirect approaches. Modern complexometric and precipitation techniques permit the direct determination of strontium using metallochromic indicators or high frequency conductometry.

Synoptic Survey

Classical Methods

Acid-Base Titration. Strontium as the oxide or carbonate is dissolved in a known excess of standard hydrochloric or nitric acid. The excess acid is then titrated with standard base [1].

Precipitation Titration. Strontium is converted to the chloride and the chloride is determined by standard procedures using either Mohr's or Fajans' method [2].

Contemporary Methods

Complexometric Titrations. Strontium is titrated photometrically in the presence of phthaleine as the metallochromic indicator [3].
Strontium is determined by titration with diethylenetrinitrilopentaacetic acid (DTPA) to an Eriochrome Black T end point [4].

Precipitation Titration. Strontium is precipitated as $SrCl_2$ from acetone solutions using standard lithium chloride as the titrant [5].

Outline of Recommended Contemporary Methods for Strontium

Complexometric Titrations

Method 1. Strontium is titrated to a spectrophotometric end point (580 mμ) with standard EDTA. Phthaleine is added as the metallochromic indicator, and the titration is carried out after adjusting the pH of the solution to 10-11.5. The titration curve shows zero slope corresponding to the absorbance of the strontium-phthaleine complex until the end point is virtually reached. Excess reagent produces a decrease in absorbance due to removal of the strontium by the EDTA from the less stable phthaleine complex. The relative error is $\pm$ 1% at strontium concentrations of 0.1-6.0 mg per 50 ml [3].

Method 2. Strontium is titrated with 0.1 M DTPA containing 0.015 M Mg-DTPA chelate. Eriochrome Black T is the preferred indicator. The relative standard deviation amounts to 0.3% or less.

Precipitation Titration

Method 1. Advantage is taken of the relative insolubility of strontium chloride in nonaqueous solvents. Strontium as the iodide in acetone is precipitated as the chloride by titration with lithium chloride dissolved in acetone. The titration is monitored by high frequency conductometry. The titration curve shows zero slope with a sharp break and change to a positive slope at the end point [5].

References

[1]. N. H. Furman, ed., *Standard Methods of Chemical Analysis*, 6th Ed., D. Van Nostrand, Princeton, N. J., Vol. 1, 1962, p. 997.

[2]. N. H. Furman, ed., *Standard Methods of Chemical Analysis*, 6th Ed., D. Van Nostrand, Princeton, N. J., Vol. 1, 1962, pp. 997-998.

[3]. K. Ogawa and S. Musha, *Bull. Univ. Osaka Perfect. Ser. A.*, *8*, 63 (1960).

[4]. E. D. Olsen and R. J. Novak, *Anal. Chem.*, *38*(1), 152-153 (1966).

[5]. G. Henroin and E. Pungor, *Anal. Chim. Acta*, *40*, 41-48 (1968).

BARIUM

Methods for the determination of barium involve complexometric and precipitation titrations. Older, but still useful, classic methods involve addition methods based on acid-base and redox titrimetry.

Synoptic Survey

Classical Methods

Acid-Base Titration. Barium carbonate is dissolved in excess standard sulfuric acid and the excess acid is titrated with standard base [1].

Redox Titration. Barium is precipitated as barium chromate. The chromate is reduced with excess iron(II) sulfate and the excess is titrated with standard potassium permanganate [2].

Complexometric Titration. Barium is titrated with standard EDTA using Eriochrome Black T as the indicator [3].

Precipitation Titrations. Barium is precipitated as barium chromate using as the titrant standard potassium dichromate [4].
Barium is titrated with standard sulfate solution with tetra-

hydroquinone as the indicator [5].

Contemporary Methods

Complexometric Titrations. Barium is titrated with standard EDTA. The end point is determined spectrophotometrically [6].
Barium is titrated with diethylenetrinitrilopentaacetic acid (DTPA) [7].

Precipitation Titrations. Barium is titrated amperometrically with lithium sulfate solution in a medium containing tetraethylammonium bromide [8].
Barium is titrated with standard lithium chloride in acetone solution [9].

Outline of Recommended Contemporary Methods for Barium

Complexometric Titrations

Method 1. Barium is readily determined by titration with standard EDTA. However, especially in the presence of the other alkaline earths, the end point is determined more accurately photometrically. Methylthymol blue and Eriochrome Black T are used as indicators in double titrations. In mixtures of barium and calcium, both metals can be determined provided that barium is the major constituent. If in addition to the calcium, strontium is also present, then barium and the sum of the calcium and strontium concentrations can be determined. Here again, the barium must be the major constituent. Magnesium interferes with the procedure. The determination is carried out at a pH of 11.1 ± 0.1 using a 650 mμ filter on the spectrophotometer [6].

Method 2. Barium is titrated with standard diethylenetrinitrilopentaacetic acid (DTPA). The method is more accurate than the usual one with EDTA. With EDTA the accuracy is limited to 1 to 2%. If DTPA is used as the titrant, the relative error is 0.3% or less. The titrant used is 0.01 M DTPA containing 0.015 M of Mg-DTPA chelate.

Precipitation Titrations

Method 1. Barium is titrated amperometrically with lithium sulfate in ethyl alcohol-water solution with tetraethylammonium bromide as the supporting electrolyte. A dropping mercury electrode is used as the indicator electrode. Since barium concentrations as low as 5×10^{-5} M may be determined, the method is of value in the analysis of radioactive samples containing small amounts of barium. The method is rapid, one titration is completed in less than half an hour. The accuracy decreases from less than 1% error for concentrated samples ($>2.5 \times 10^{-3}$ M) to less than 10% error for more dilute samples ($<5 \times 10^{-5}$ M) [8].

Method 2. Barium together with sodium or potassium dissolved in acetone is titrated with standard lithium chloride in acetone. This new method is based on the changed conditions of solubility of the respective barium, sodium, and potassium salt in nonaqueous media. Both acetone and methyl isobutyl ketone are recommended as solvents, but the acetone is the preferred medium because the precipitation equilibrium is reached very quickly. The end point is determined oscillometrically. The error varies but is usually within $\pm$ 1% [9].

References

[1]. N. H. Furman, ed., *Standard Methods of Chemical Analysis*, 6th Ed., D. Van Nostrand, Vol. 1, Princeton, N. J., 1962, p. 150.

[2]. N. H. Furman, ed., *Standard Methods of Chemical Analysis*, 6th Ed., D. Van Nostrand, Vol. 1, Princeton, N. J., 1962, p. 150.

[3]. T. J. Marms, M. U. Reschovsky, and H. F. Carta, *Anal. Chem.*, *24*, 908 (1953).

[4]. N. H. Furman, ed., *Standard Methods of Chemical Analysis*, 6th Ed., D. Van Nostrand, Vol. 1, Princeton, N. J., 1962, p. 149.

[5]. N. H. Furman, ed., *Standard Methods of Chemical Analysis*, 6th Ed., D. Van Nostrand, Vol. 1, Princeton, N. J., 1962, p. 150.

[6]. H. F. Combs and E. L. Grove, *Anal. Chem.*, *36*(2), 400-402 (1964).

[7]. E. D. Olsen and R. J. Novak, *Anal. Chem.*, *38*(1) 152-153 (1966).

[8]. H. E. Zittel, F. J. Miller, and P. F. Thomason, *Anal. Chem.*, *31*(8), 1351-1353 (1959).

[9]. G. Henrion and E. Pungor, *Anal. Chim. Acta*, *40*, 41-48 (1968).

ZINC

Methods for the determination of zinc include complexometric and precipitation titrations.

Synoptic Survey

Classical Methods

Precipitation Titration. Zinc is titrated with standard potassium ferrocyanide solution using uranium nitrate as an external indicator or diphenylamine as an internal indicator [1].

Contemporary Methods

Complexometric Titrations. Zinc is determined by a weight titration with standard EDTA. Eriochrome Black T is the indicator. The end point is found photometrically [2].

Zinc and cadmium are determined in the same sample by titration with standard ethylene glycol bis(2-aminoethylether)tetraacetic acid (EGTA). Murexide is used as the indicator, the end point is determined photometrically [3].

Zinc is titrated amperometrically with standard 1,2-cyclohexanedinitrilotetraacetic acid (CDTA) in 2-propanol [4].

Zinc and cadmium in the same solution may be determined in the presence of copper by masking the copper and cadmium with 2-mercaptopropionic acid (MPA). The zinc is titrated with triethylenetetraminehexaacetic acid (TTHA) [5].

Precipitation Titration. Zinc is titrated amperometrically with potassium tetracyano-mono-(1,10-phenanthroline)-ferrate (II) [6].

Outline of Recommended Contemporary Methods for Zinc

Complexometric Titrations

Method 1. Zinc is determined by weight titration with standard EDTA using Eriochrome Black T as the indicator. The end point is found photometrically. The method is especially suitable for the accurate determinations of very small amounts of zinc. The standard error is about 0.02% for 0.6 mg of zinc and about 0.1 for solutions containing 0.06 mg of zinc [2].

Method 2. Zinc and cadmium can be titrated in one sample with EGTA. The determination is carried out at an adjusted pH of 10 using murexide as the indicator. The end point is determined photometrically at about 450 mμ. The two breaks in the titration curve correspond to the values for cadmium and zinc. Accuracy and precision are satisfactory [3].

Method 3. Zinc is titrated in nonaqueous solutions amperometrically with standard CDTA. The preferred solvent is 2-propanol. The titration is carried out in the presence of a large excess of base which is added to neutralize the protons released in the formation of the zinc chelate, and to stabilize the environment. The titrant, CDTA, which is usually 0.1 M is dissolved in 95% ethanol which is 1 M with ethanolamine. The average deviation is 1×10^{-5} M for zinc concentrations of 1.45×10^{-3} M [4].

Method 4. Zinc in the presence of cadmium and copper may be determined by first masking the cadmium and copper with MPA. The zinc is then titrated with standard solutions of TTHA. If an excess of CDTA is then added and the excess back-titrated with standard zinc nitrate solutions, values for the amounts of cadmium are obtained. The determination is run in slightly acidic solutions to prevent interferences from the alkaline earth metals. Xylenol orange is the preferred indicator. The precision is satisfactory [5].

Precipitation Titration

Zinc is titrated amperometrically with standard potassium tetra-cyano-mono-(1,10-phenanthroline)-ferrate(II). This titrant is to be preferred to potassium hexacyanoferrate(II) because of the increased accuracy in the determination of small amounts of zinc. The lower limit for zinc concentrations is approximately 8×10^{-6} M; the smallest zinc concentration is 0.0124 mg of zinc per 25 ml. Relatively large amounts (ca. 0.1 M) of magnesium or aluminum and moderate amounts (ca. 0.02 M) of the alkaline earth metals do not interfere. The titration time is considerably shorter, and strict adherence to the prescribed experimental conditions necessary in the titration of zinc with potass-ium hexacyanoferrate(II) is no longer necessary. Iron(II), cobalt(II), copper(II), and cadmium(II) are quantitatively precipitated by the re-agent. Interferences in the titration due to Ag^+ or Hg_2^{2+} are prevented by prior precipitations as the chlorides. The relative standard dev-iation is 0.36% for 2.50 mg of zinc and 1.8% for 0.250 mg of zinc [6].

References

[1]. N. H. Furman, ed., *Standard Methods of Chemical Analysis*, 6th Ed., D. Van Nostrand, Vol. 1, Princeton, N. J., 1962, pp. 1235-1239.

[2]. D. B. Scaif, *Analyst*, *88*, 618-621 (1963).

[3]. H. Flaschka and T. B. Carley, *Talanta*, *11*, 423-431 (1964).

[4]. P. Arthus and B. R. Hunt, *Anal. Chem.*, *39*(1), 95-97 (1967).

[5]. R. Přibil and V. Vesely, *Talanta*, *12*, 475-478 (1965).

[6]. A. A. Schilt and A. V. Nowak, *Anal. Chem.*, *36*(4), 485-488 (1964).

CADMIUM

Methods for the determination of cadmium include redox and complexometric titrations.

Synoptic Survey

Classical Methods

Redox Titration. Cadmium is precipitated as the sulfide which is dissolved in hydrochloric acid, and the equivalent amount of sulfur is titrated with standard iodine solution [1].

Contemporary Methods

Complexometric Titrations. Cadmium is titrated potentiometric-ally with standard tetraethylenepentamine (Tetren) [2].

Cadmium is determined by weight titration with standard EDTA and
Eriochrome Black T as the indicator [3].

Cadmium in the presence of zinc, masked with sodium hydroxide,
is titrated with standard ethylene-bis(oxyethylenenitrilo)tetraacetic
acid (EGTA) [4].

Cadmium is titrated amperometrically with standard 1,2-cyclo-
hexanedinitrilotetraacetic acid, (CDTA), in 2-propanol [5].

Outline of Recommended Contemporary Methods for Cadmium

Complexometric Titrations

Method 1. Cadmium is titrated potentiometrically with standard
Tetren which is a much better selective titrant than Trien. Copper,
calcium, magnesium, aluminum, lanthanum, or manganese do not interfere. The deviation for 25.12 mg of cadmium in the presence of calcium,
magnesium, aluminum, or lanthanum is $\pm$ 0.4%. For 12.48 mg of cadmium
in the presence of 6.36 mg of copper, the standard deviation is $\pm$ 0.4%
[2].

Method 2. Cadmium is determined by a weight titration with standard EDTA using Eriochrome Black T as the indicator. The use of weight
burettes as well as photometric end-point detection increases the sensitivity of the method. Very small amounts of cadmium can be determined. The standard error for samples containing 0.6 mg of cadmium is
0.02%. For samples containing 0.06 mg of cadmium, the standard error
is 0.1% [3].

Method 3. Both cadmium and zinc form complexes with EGTA. However, the two complexes differ only slightly in stability.

$$Cd-EGTA \quad chelate \quad log \; k = 14.5$$

$$Zn-EGTA \quad chelate \quad log \; k = 16.7$$

In sodium hydroxide solutions, zinc-EGTA chelate is unstable. Zincate
ion is formed while the Cd-EGTA chelate is stable. Advantage is taken
of this difference to titrate cadmium in the presence of zinc. An
excess of standard EGTA is added, the zinc is masked with sodium hydroxide, and the excess EGTA is titrated with standard calcium chloride
solution. Metalphthalein is used as the indicator. Lead, if present,
behaves like zinc forming the plumbate ion. Plumbate and zincate ions
form colorless to faintly pink colored complexes with metalphthalein
in contrast to calcium which gives an intense red-violet color. Interfering ions such as iron and aluminum are eliminated by masking with
triethanolamine. The method is used for the determination of cadmium
in crude zinc metal, and in common cadmium pigments. The precision
and accuracy are excellent [4].

Method 4. Cadmium is titrated amperometrically in 2-propanol
with CDTA. During the titration, the protons released from the CDTA
must be neutralized by the addition of a large excess of base. The
presence of an excess of base also helps promote chelation. The

titrant used is 0.1 M with CDTA and 1 M with ethanolamine in 95% ethanol. The average deviation in the determination of samples containing 1.5 x 10^{-3} M cadmium is 7 parts per thousand [5].

References

[1]. N. H. Furman, ed., *Standard Methods of Chemical Analysis*, 6th Ed., D. Van Nostrand, Princeton, N. J., Vol. 1, 1962, pp. 253-254.

[2]. C. N. Reilly and A. Vavonlis, *Anal. Chem.*, *31*(2), 243-248 (1959).

[3]. D. B. Scaif, *Analyst*, *88*, 618-621 (1963).

[4]. R. Přibil and V. Vesely, *Chemist-Analyst*, *55*(1), 4-5 (1967).

[5]. P. Arthur and B. R. Hunt, *Anal. Chem.*, *39*(1), 95-97 (1967).

MERCURY(I)

The primary method for the determination of mercury(I) is by redox titrimetry.

Synoptic Survey

Classical Methods

Redox Titration. Mercury(I) is titrated with standard ceric sulfate [1], or standard permanganate [2].

Contemporary Methods

Mercury(I) is titrated with standard ceric sulfate in 4 N hydrochloric acid with iodine monochloride as the catalyst [3].

Mercury(I) is titrated with standard potassium dichromate and iodine monochloride as the catalyst [4].

Mercury(I) is allowed to react with standard ferricyanide and the resulting ferrocyanide is titrated with standard ceric sulfate [5].

Outline of Recommended Contemporary Methods for Mercury(I)

Redox Titrations

Method 1. Mercury(I) can be titrated by an improvement on the classic method by carrying out the titration in approximately 4 N hydrochloride in the presence of iodine monochloride as a catalyst. Several advantages result from the use of this catalyst. Not only is a somewhat sluggish reaction speeded up, but it is no longer necessary to heat the reaction mixture and the application of an empirical factor is superfluous. Results are good [3].

Method 2. Mercury(I) is titrated with standard potassium dichromate solution. The slow reaction is speeded up by the addition of iodine monochloride. The titration is carried out in hydrochloric acid which must be at least 7.5 N. Results are satisfactory [4].

Method 3. Mercury(I) is titrated by the addition of an excess of standard ferricyanide solution in the presence of cyanide ion. The equivalent amount of ferrocyanide formed is determined by titration with standard ceric sulfate using phenanthroline as the indicator. The standard deviation is 0.2% [5].

References

[1]. H. H. Willard and P. Young, *J. Am. Chem. Soc.*, *52*, 557 (1930).

[2]. N. H. Furman, ed., *Standard Methods of Chemical Analysis*, 6th Ed., D. Van Nostrand, Princeton, N. J., Vol. 1, 1962, p. 663.

[3]. G. J. Rao and K. B. Rao, *Z. Anal. Chem.*, *168*, 81-83 (1959).

[4]. G. J. Rao and K. B. Rao, *Z. Anal. Chem.*, *169*, 247-248 (1960).

[5]. F. Lucena-Conde, V. S. Perez, and A. M. Beato, *Stud. Chem. Univ. Salamanca 1966*(2), 75-77.

MERCURY(II)

Methods for the determination of mercury(II) include redox, complexometric and precipitation titrimetry.

Synoptic Survey

Classical Methods

Precipitation Titrations. Mercury(II) is titrated with standard potassium iodide using starch as the indicator (Seamon's method).
Mercury(II) is separated from solution as the sulfide which is dissolved and titrated with standard thiocyanate using a ferric ion indicator [2].

Contemporary Methods

Redox Titrations. Mercury(II) is precipitated as the para-periodate. The latter is dissolved in hydrochloric acid and titrated iodometrically [3].

Mercury(II) is titrated with sulfhydryl groups which are internally generated by constant current titrimetry from the stable mercury complex of thioglycolic acid [4].

Mercury(II) is titrated with internally generated sulfhydryl groups from the mercury(II) complex of monothioethyleneglycol (MTEG) [5].

Complexometric Titration. Mercury(II) is determined by titration with standard potassium iodide solution in the presence of bismuth ions and cyclohexanone [6].

Precipitation Titrations. Mercury(II) is determined by high-frequency titration with standard thiocyanate [7].

Mercury(II) is titrated by coulometrically generated iodide [8].

Outline of Recommended Contemporary Methods for Mercury(II)

Redox Titrations

Method 1. Mercury(II) is precipitated as the mercuric paraperiodate. The precipitate is dissolved in hydrochloric acid; benzene and potassium iodide are added, and the liberated iodine is titrated with standard sodium thiosulfate. Copper, cadmium, aluminum, or nickel do not interfere if their concentrations are ≤ 1.5 times the concentration of the mercury(II). The error in the determination is $\pm 0.2\%$ [4].

Method 2. Mercury(II) is determined by titration with the sulfhydryl group obtained by constant current titrimetry of the stable mercury complex of thioglycolic acid, $HSCH_2COOH$. The sulfhydryl group shows a great affinity for metals, and is also a strong reducing agent in the upper pH range. For ordinary titrimetry organic sulfhydryl compounds tend to be unstable towards air oxidation, have objectionable odors, and may show excessive volatility and limited solubility. The internal generation of such groups from stable compounds by constant current titrimetry overcomes the difficulties usually met in employing such compounds. The end point may be found by means of a mercury-indicating pM electrode or by amperometry at two indicator mercury electrodes. The accuracy in the determination is $\pm 0.5\%$ at the milligram level [4].

Method 3. Mercury(II) is titrated with organic sulfhydryl groups obtained from the mercury complex of monothioethylene glycol (MTEG). Constant current coulometric generation of the sulfhydryl group is feasible. Interferences due to the formation of secondary complexes including the solubilizing hydroxyl group are negligible. The sensitivity is higher than that found when thioglycolic acid is used as the source of the sulfhydryl groups [5].

Complexometric Titration

Mercury(II) can be titrated in acidic solution containing bismuth(III) with potassium iodide. The HgI_3^{-1} is extracted from the

aqueous solution by the addition of cyclohexanone. When all the mercury is complexed, the bismuth combines with the KI forming a compound
which colors the cyclohexanone orange-red. Chloride, bromide, and most
heavy metals do not interfere; Tl(I), Ag(I), and thiocyanate ion do
interfere in the determination. In the case of iron(III) prior reduction with hydroxylamine is advised. The precision and accuracy are
very good even for small amounts of mercury. For samples containing
10.03 mg of mercury the standard deviation is $\pm$ 0.11 mg and the vari-
ance is 1% [6].

Precipitation Titrations

Method 1. Mercury(II) is determined by high-frequency titration
with standard thiocyanate solution. The inflection in the titration
curve is due to the formation of mercury(II) thiocyanate. The method
is well suited for the accurate determination of very small amounts
of mercury (0.63 to 12.6 mg) in highly dilute solutions. The method
is rapid and accurate. Nitric acid may be present in concentrations
up to 0.03 N. Even at extreme dilutions (1/16000 to 1/32000 M) the
error is 2% and 4% respectively [7].

Method 2. Mercury(II) is titrated coulometrically with iodide.
The end point is indicated by a silver/silver iodide electrode. The
electrolyte contains perchloric acid. It is also possible to determine both mercury(I) and mercury(II) in a mixture by oxidizing the
mercury(I) to mercury(II) with the use of constant potentiometric
coulometry. These data and the data one obtains from the titration of
the mixture prior to oxidation of the mercury(I) immediately give values for the amounts of both mercury(I) and mercury(II) [8].

References

[1]. N. H. Furman, ed., *Standard Methods of Chemical Analysis*, 6th
       Ed., D. Van Nostrand, Princeton, N. J., Vol. 1, 1962, p. 661.

[2]. N. H. Furman, ed., *Standard Methods of Chemical Analysis*, 6th
       Ed., D. Van Nostrand, Princeton, N. J., Vol. 1, 1962, p. 662.

[3]. A. A. Verdizade, *Uch. Zap. Azerb. Gos. Univ., Ser. Khim. Nauk.*,
       2, 77-80 (1965).

[4]. B. Miller and D. N. Hume, *Anal. Chem.*, 32(4), 524-530 (1960).

[5]. B. Miller and D. N. Hume, *Anal. Chem.*, 32(7), 764-769 (1960).

[6]. E. Jackwerth and H. Specker, *Z. Anal. Chem.*, 167, 268-271 (1959).

[7]. F. Magno, *Anal. Chim. Acta*, 40(3), 431-435 (1968).

GROUP III

	Lanthanide Rare Earths	Actinides
Boron	Lanthanum	Thorium
Aluminum	Cerium	Uranium
Gallium	Neodyminum	Neptunium
Indium	Samarium	Plutonium
Thallium	Europium	
Scandium	Gadolinium	
Yttrium	Terbium	
	Dysprosium	
	Thulium	
	Ytterbium	
	Lutetium	

BORON

Methods for the determination of boron include redox and precipitation titrations.

Synoptic Survey

Classical Methods

Acid-Base Titration. Boron in solution as boric acid is titrated in the presence of a polyhydric alcohol such as glycerol or mannitol with a standard base and phenolphthalein as the indicator [1].

Contemporary Methods

Acid-Base Titrations. Boric acid is titrated thermometrically with standard NaOH [2].
Boric acid is titrated amperometrically with standard fructose in 0.1 M lithium chloride [3].

Redox Titration. Boron as the boranate (borohydride) is oxidized to metaborate by excess standard hypochlorite in strongly basic solution. The excess is determined iodometrically [4].

Precipitation Titration. Tetrafluoborate is titrated in methylene chloride with standard tetraphenylarsonium chloride [5].

Outline of Recommended Contemporary Methods for Boron

Acid-Base Titrations

Method 1. Orthoboric acid is determined by thermometric titration with standard sodium hydroxide. The end point is detected by the measurement of the heat evolved during the neutralization. However, the concentration of the titrant must be 100 times greater than the concentration of the unknown acid sample in order to minimize changes in heat capacity. Optimum concentration of orthoboric acid is 10^{-3} to 10^{-4} M. The method is simple and rapid [2].

Method 2. Aqueous borate solutions are titrated amperometrically with standard fructose in a 0.1 M lithium chloride-0.1 M lithium hydroxide solution. The depression of the polarographic wave of fructose in the LiCl-LiOH solution is proportional to the concentration of borate ions added. The applied voltage is -2.05 V. At this voltage, all polarographically active cations interfere. These interferences are removed by passage through the hydrogen form of a cation exchange resin. Lithium, chloride, iodide, bromide, phosphate, and carbonate do not interfere. The interference of bromates and iodates is removed by prior treatment of the sample with sulfur dioxide in an alkaline solution. The accuracy is about $\pm$ 0.01 mg for samples containing 0.5-2.00 mg of boron in 25 ml of solution [3].

Redox Titration

Boranate (borohydride) ion is quantitatively oxidized by an excess of hypochlorite ions in strongly basic solutions.

$$BH_4^- + 4\ OCl^- \rightarrow BO_2^- + 4\ Cl^- + 2\ H_2O$$

An excess of the oxidizing agent is added, the solution is acidified and potassium iodide is added. The liberated iodine is then titrated with standard sodium thiosulfate [4].

Precipitation Titration

Small amounts of tetrafluoborate are determined by partition titration. The procedure may be carried out in the presence of borate and fluoride. Methylene chloride is used as the solvent because of its high distribution coefficient.

The tetrafluoborate is titrated with standard tetraphenylarsonium chloride which acts not only as a precipitating agent but also as an extractant. The end point may be determined either photometrically or amperometrically. Titration curves obtained by either of these two methods coincide at high concentrations. Anions such as nitrate, chlorate, perrhenate, etc. are extracted by tetraphenylarsonium chloride and must be absent. The standard deviation for samples containing 51.7 - 1068.5 µg is 0.6% [5].

References

[1]. N. H. Furman, ed., *Standard Methods of Chemical Analysis*, 6th Ed., D. Van Nostrand, Princeton, N. J., 1962, Vol. 1, p. 215.

[2]. F. Pechar, *Chem. Listy*, *59*(9), 1073-1075 (1965).

[3]. W. B. Swann, W. M. McNoble, and J. F. Hazel, *Anal. Chim. Acta*, *22*, 76-81 (1960).

[4]. C. Z. Harzdorf, *Z. Anal. Chem.*, *210*(1), 12-17 (1965).

[5]. K. Behrends, *Z. Anal. Chem.*, *216*(1), 13-20 (1966).

ALUMINUM

Methods for the determination of aluminum include acid-base and complexometric titrations.

Synoptic Survey

Classical Methods

Acid-Base Titration. Aluminum is determined acidimetrically in accordance with the reactions:

$$Al^{+3} + 3\ OH^- \rightarrow Al(OH)_3$$

$$Al(OH)_3 + 6\ KF \rightarrow AlF_3 \cdot 3\ KF + 3\ KOH$$

The aluminum is sequestered by the addition of KF. The liberated KOH is titrated with standard alkali [1].

Contemporary Methods

Complexometric Titrations. Aluminum is titrated with 0.2 M disodium barium ethylenedinitrilotetraacetic acid using pyrocatechol violet as the indicator [2].

Aluminum is titrated with standard 1,2-cyclohexylenedinitrilotetraacetic acid (CDTA) and xylenol orange as the indicator [3].

Aluminum and iron in a mixture are titrated consecutively with standard hydroxyethylethylenediaminetriacetic acid (HEDTA). The indicator used is 8-hydroxy-7-[(4-sulfo-1-naphthyl)-azo]-5-quinoline sulfonic acid (SNAZOXS).

Outline of Recommended Contemporary Methods for Aluminum

Complexometric Titrations

Method 1. Micro amounts of aluminum in the form of the nitrate
are titrated with 0.2 M disodium barium ethylenedinitrilotetraacetate
in the presence of sulfate at pH 3-4. The preferred indicator is
pyrocatechol violet. For samples containing 0.04-4.17 mg of aluminum,
the error varies between 0.10 and 0.50% [2].

Method 2. Milligram amounts of aluminum are titrated with standard CDTA. In many cases, since the titrant is not very selective, it
is first separated from other interfering elements. This is especially
true during the analysis of modern aluminum alloys. Many times a mercury cathode will be sufficient to remove most interfering elements.
For alloys containing titanium and zirconium, a prior extraction is
carried out using tri-n-octylphosphine oxide (TOPO). Manganese is removed by oxidation with chlorate and filtration of the resulting manganese dioxide. In the actual titration, an excess of CDTA is added and
the excess is back titrated with xylenol orange as the indicator, and
standard zinc solution. The method is fairly rapid. The mean absolute error is less than 0.02% for an alloy containing about 3% aluminum [3].

Method 3. In a mixture of aluminum and iron, the latter is first
titrated in nitric acid solution at pH 2 with standard HEDTA using
SNAZOXS as the indicator. At the end point, the solution is deep red
in color. An excess of standard HEDTA is then added, and after adjusting the pH to 5, the excess is back-titrated with standard copper solution. The iron titration must be performed at 50° to increase the
speed of a rather sluggish reaction. The relative standard deviations
in the analysis of a mixture containing 0.1 mmole of both iron and
aluminum are 0.6 and 0.8% respectively. The relative errors are -0.3
and +0.4% respectively [4].

References

[1]. N. H. Furman, ed., *Standard Methods of Chemical Analysis*, 6th
      Ed., D. Van Nostrand, Princeton, N. J., 1962, Vol. 1, pp. 650-
      651.

[2]. G. Asensi Morva, *Anales Real. Soc. Espan. Fis. Quim. (Madrid)
      Ser. B.*, *58*, 523-530 (1962).

[3]. K. E. Burke and M. C. Davis, *Anal. Chem.*, *36*(1), 172-175 (1964).

[4]. D. A. Aikens and F. J. Bahbah, *Anal. Chem.*, *39*(6), 646-649
      (1967).

GALLIUM

Methods for the determination of gallium include complexometric and precipitation titrations.

Synoptic Survey

Classical Methods

Precipitation Titration. Gallium is titrated with standard potassium ferrocyanide either visually or potentiometrically [1,2].

Contemporary Methods

Complexometric Titrations. Gallium is titrated with standard EDTA using a copper-PAN indicator [3].

Gallium is titrated indirectly with standard EDTA. The titration is carried out using a metalfluorochromic indicator such as calceine blue in ultraviolet light.

Gallium is titrated with standard EDTA at 365 mμ using Morin (2',3,4',5,7-pentahydroxyflavone) as the indicator [5].

Gallium is precipitated with diantipyrinylpropylmethane in 6 N hydrochloric acid. The precipitate is dissolved in ammonium acetate and titrated with standard EDTA using as an indicator 1-(2-pyridylazo)-2-naphthol (PAN) [6].

Precipitation Titrations. Gallium is titrated amperometrically with standard cupferron solution and a vibrating electrode [7].

Gallium is titrated amperometrically with a standard potassium ferrocyanide solution using a rotating platinum electrode [8].

Outline of Recommended Contemporary Methods for Gallium

Complexometric Titrations

Method 1. Gallium in the presence of aluminum is titrated in a boiling buffered solution at a pH of 1.6-2.0 with standard EDTA. A good indicator is available in the copper-PAN system. Aluminum is masked with fluoride since this complex is more stable than the corresponding gallium complex. In contrast, the gallium-EDTA complex is more stable than the aluminum-EDTA complex. At the end point the color of the solution changes from violet to yellow. In the analysis of various samples containing 1 mole of gallium per 0.67 mole of aluminum, the recovery was 99.98 $\pm$ 0.84% (at the 95% confidence level) in a range of 99.4 to 100.5% [3].

Method 2. Gallium is titrated indirectly with excess standard EDTA. The use of an indicator such as calceine blue, a metalfluorochromic indicator, greatly improves the method. The excess EDTA is back-titrated with 0.025 M copper sulfate solution preferably in an ultraviolet titration assembly with a cadmium sulfide detector cell and a microammeter. During the titration, the initial fluorescence decreases rapidly and then more slowly to the end point where a sharp drop in the fluorescence occurs. The method is fast, precise, and accurate. For samples containing 10 mg of gallium, the accuracy is approximately 0.1% and the relative standard deviation is about 0.4% [4].

Method 3. Gallium is titrated with a standard solution of EDTA (M $\leq$ 0.05 M). The titration is carried out at pH 2.5-4.5 at 365 mμ using Morin as the indicator. During the titration the fluorescence of the Morin decreases and disappears entirely at the end point. In the presence of interfering metals, gallium is extracted from the solution prior to titration. The extraction is made from a solution acidified with hydrochloric acid (1:1) using n-butyl acetate. The relative error varies between 0.7 and 4.8%. The method is quite sensitive and may be used to determine gallium in samples containing 1-10 γ of gallium in 2 ml of solution [5].

Method 4. Gallium reacts with diantipyrinylpropylmethane in 6 N hydrochloric acid to form a precipitate. This precipitate is soluble in ammonium acetate solutions containing tartaric acid at pH 5. The gallium in such solutions is then titrated with standard EDTA with PAN as the indicator. Zinc, cadmium, copper, aluminum, nickel, manganese, mercury, indium, cobalt, and bismuth do not interfere. Interferences are noted for iron(III) and thallium. Interference due to iron(III) is removed by prior reduction [6].

Precipitation Titrations

Method 1. Gallium is titrated amperometrically with a standard cupferron solution. Citric and tartaric acids do not interfere. In the presence of oxalic and lactic acids the results tend to be low. In solutions containing iron, the iron should be reduced with metallic cadmium or titanous chloride and the gallium extracted with n-butyl acetate from the solution made 6 N with respect to hydrochloric acid. The gallium is then extracted with water from the butyl acetate solution, the water extract acidified with sulfuric acid to a pH 3-4 and the gallium then titrated with standard cupferron in the presence of sodium chloride. Aluminum may be present up to 50-fold and zinc up to 200-fold the amount of gallium without interference [7].

Method 2. Gallium is titrated amperometrically with standard potassium ferrocyanide. A rotating platinum electrode is used and the determination is carried out in 0.1-0.5 M hydrochloric acid and 0.1 M potassium chloride as the supporting electrolyte. The applied potential is +0.9 V. The method is rapid and of practical importance in the analysis of aluminum-containing rocks [8].

References

[1]. H. D. Kirschmann and T. B. Ramsey, *J. Am. Chem. Soc.*, *50*, 1632-1635 (1928).

[2]. R. Belcher, A. J. Nutten, and W. I. Stephen, *J. Chem. Soc.*, *1952*, 2438-2439.

[3]. J. S. Mee and J. D. Corbott, *Chemist-Analyst*, *50*, 74-76 (1961).

[4]. H. N. Elsheimer, *Talanta*, *14*(1), 97-102 (1967).

[5]. L. A. Solovera, K. P. Stolyarov, and N. N. Grigorlev, *Vestn. Leningr. Univ., Ser. Fiz. i Khim.*, *19*(16), No. 3, 134-139 (1964).

[6]. A. I. Busev and V. G. Tiptsova, *Zhur. Anal. Khim.*, *15*, 698-700 (1960).

[7]. I. A. Tserkovnitskaya, A. I. Kallinin, and Yn. V. Movachevskii, *Zavodsk. Lab. 26*, 797 (1960).

[8]. I. L. Bagbanly and S. I. Bagbanly, *Dokl. Akad. Nauk. Azerb. SSR 23*(1), 13-17 (1967).

INDIUM

Methods for determination of indium include complexometric and precipitation titrations.

Synoptic Survey

Classical Methods

Precipitation Titration. Indium is titrated with standard potassium ferrocyanide [1].

Contemporary Methods

Complexometric Titrations. Indium is titrated with 0.05 M EDTA with xylenol orange as the indicator [2].
Indium is titrated by adding an excess of standard EDTA and back-titrating the excess with standard mercuric nitrate solution [3].
Indium is titrated amperometrically with standard EDTA [4].

Precipitation Titration. Indium is titrated with 0.03 M sodium diethyldithiocarbamate (Na-DDTC) using pyrocatechol violet as the indicator. Instead of this visual method, the end point may also be determined amperometrically or potentiometrically [5].

Outline of Recommended Contemporary Methods for Indium

Complexometric Titrations

Method 1. Indium is titrated at an elevated temperature of 50-60° with 0.05 M EDTA and xylenol orange as the indicator. The titration is carried out at a pH 3.0-3.5. Iron(III), aluminum, bismuth, lead, thorium, titanium, and gallium interfere. The interference by cadmium, zinc, copper, cobalt, nickel, manganese, and the uranyl ion is prevented by masking with a 2-3-fold excess of phenanthroline. Thallium, molybdenum, tungsten, chromium, vanadium, and silver do not interfere. The mean error is less than 1.5% [2].

Method 2. Indium is determined by adding an excess of standard EDTA and back-titrating potentiometrically with standard mercuric nitrate at pH 10. Mixtures of indium with various metals can often be analyzed by masking the interfering ions with cyanide, titrating directly with EDTA, and following this by back-titrating a second sample with mercuric nitrate. The method is simple and rapid. Samples containing 114 μg - 18 mg of indium can be analyzed with high accuracy and precision [3].

Method 3. Indium is titrated amperometrically with standard EDTA at a rotating tantalum microanode at a pH $\geq$ 1 and a potential of 1.2-1.4 V. Large concentrations of magnesium, calcium, strontium, barium, beryllium, aluminum, cobalt, zinc, silver, lead, and uranyl ion do not interfere, nor do moderate amounts of nickel, chromium(III), cerium(IV), tin(IV), and mercury complexed with chloride ion. Strong interferences are shown by iron(III), copper, and antimony(III) [4].

Precipitation Titration

Samples containing 2-12 mg indium are titrated with 0.03 M NaDDTC at pH 4 using pyrocatechol violet as the indicator. The end point may also be determined potentiometrically with a silver indicator electrode. Smaller amounts of indium in the range of 0.02-2.00 mg are determined by amperometric titration at -0.8 V in an acetate buffer at a pH of 4.4 by measuring the decrease of the cathodic wave of indium(III) [5].

References

[1]. N. B. Bray and H. D. Kirschman, *J. Am. Chem. Soc.*, *49*, 2739 (1927).

[2]. M. Kopanica and R. Pribil, *Collection Czechoslov. Chem. Communs.* *25*, 2230-2232 (1960).

[3]. H. Khalifa and M. M. Khater, *Z. Anal. Chem.*, *184*, 92-98 (1961).

[4]. V. A. Khadeev and I. Bezarov, *Uzbeksk Khim. Zh.*, *6*, No. 5, 47-53 (1962).

[5]. R. Staroscik and H. Siaglo, *Chem. Anal. (Warsaw)* 10(2), 265-270 (1965).

THALLIUM(I)

Methods for the determination of thallium(I) include redox and precipitation titrations.

Synoptic Survey

Classical Methods

Redox Titration. Thallium(I) is titrated with standard potassium permanganate [1], cerium(IV) [2], or potassium bromate [3].

Contemporary Methods

Redox Titrations. Thallium(I) is titrated with standard potassium dichromate in the presence of iodine monochloride and carbon tetrachloride [4].

Thallium(I) is titrated amperometrically with standard ceric sulfate or potassium dichromate in 10 N sulfuric acid with manganese(II) sulfate as a catalyst [5].

Thallium(I) is titrated in concentrated hydrochloric acid solutions with standard potassium dichromate and Siloxene as a chemiluminescent indicator [6].

Precipitation Titration. Thallium(I) is determined by high frequency titration with potassium chromate [7].

Outline of Recommended Contemporary Methods for Thallium(I)

Redox Titrations

Method 1. Thallium(I) is titrated in at least 7.5 N hydrochloric acid solution with standard potassium dichromate in the presence of iodine monochloride. The hydrochloric acid concentration must be at least 7.5 N at the end point. The titration is carried out at room temperature in a two-phase system consisting of the aqueous sample, the iodine monochloride and sufficient carbon tetrachloride to dissolve the ICl. Efficient shaking of the mixture is required at all times during the titration. At the end point the carbon tetrachloride layer changes from purple to colorless [4].

Method 2. Thallium(I) is titrated amperometrically with cerium (IV) sulfate at constant voltage. Manganese(II) sulfate or iodine monochloride acts as an effective catalyst. The titration may also be carried out using standard potassium dichromate in 10 N sulfuric acid with a manganese(II) sulfate catalyst [5].

Method 3. Thallium(I) is titrated in high concentrations of hydrochloric acid with standard potassium dichromate using the chemiluminescent indicator, Siloxene. The Siloxene is prepared by boiling calcium silicide with hydrochloric acid, filtering, and drying at room temperature. The end point is shown by the development of an orange-red chemiluminescence upon the addition of the slightest excess of the titrant. Accuracy is 0.24% and the standard deviation is $\pm$ 0.09% [6].

Precipitation Titration

Thallium(I) is determined by high frequency titration with standard potassium chromate (ca. 0.03 M) at a pH of 6.0-6.5. The method is well-suited to the determination of samples containing up to 50 mg of thallium. The accuracy is about $\pm$ 1% [7].

References

[1]. R. A. Beale, A. W. Hutchinson, and G. C. Chandler, *Ind. Eng. Chem., Anal. Ed., 13,* 240 (1941).

[2]. H. H. Willard and P. Young, *J. Am. Chem. Soc., 52,* 36 (1930).

[3]. E. Zintl and G. Rienaecker, *Z. Anorg. Chem., 153,* 276 (1926).

[4]. K. B. Rao, *Z. Anal. Chem., 165,* 193-195 (1959).

[5]. D. Singh and N. Bhatnagar, *J. Sci. Res. Banaras Hindu Univ., 15* (2), 263-268 (1964-1965).

[6]. I. Buzás and L. Erdey, *Talanta, 10,* 647-670 (1963).

[7]. A. K. Majumdar and B. K. Mitra, *Anal. Chim. Acta, 21,* 29-32 (1959).

THALLIUM(III)

Methods for the determination of thallium(III) include redox and complexometric titrations.

Synoptic Survey

Classical Methods

None of importance.

Contemporary Methods

Redox Titrations. Thallium(III) is titrated amperometrically with standard ascorbic acid at a rotating platinum electrode [1].
Thallium(III) is precipitated as the periodate which is dissolved and titrated iodometrically [2].
Thallium(III) is titrated potentiometrically or amperometrically with standard solutions of thiourea [3].
Thallium(III) chloride is titrated with standard tin(II) chloride either potentiometrically [4] or visually using KI and starch indicator [5].

Complexometric Titrations. Thallium(III) is titrated with standard EDTA and iron(III)-sulfosalicylate complex as the indicator [6].

Thallium(III) is determined in the presence of thallium(I) by EDTA titration [7].

Outline of Recommended Contemporary Methods for Thallium(III)

Redox Titrations

Method 1. Thallium(III) is titrated amperometrically with standard ascorbic acid at a rotating platinum electrode at 0.8-1.0 V using a standard calomel reference electrode. It is advantageous to measure the oxidation current of the reagent after the end point is passed. As little as 0.001 mg of thallium per milliliter can be determined. Calcium, aluminum, chromium, manganese, zinc, cobalt, nickel, uranyl ion, beryllium, titanium(IV), cerium(III), cadmium, bismuth, copper, nitrate, fluoride, phosphate, arsenate, selenate, sulfate, acetate, and molybdate do not interfere. Tartrate and citrate interfere only at concentrations greater than five times the concentration of thallium; chloride interferes only at concentrations greater than twenty times the thallium concentration. Complexon III, antimony, and mercury interfere even at low concentrations. The determination is carried out in 0.3-0.5 N sulfuric acid. An ascending type titration curve is obtained with an inflection at the end point. Precision and accuracy are good [1].

Method 2. Thallium(III) is determined by precipitation in 9-10 N sulfuric acid of the periodate by the addition of potassium periodate. The reddish-brown crystalline precipitate has the composition $Tl_5(IO_6)_3$, and is stable up to 120°. After filtration, this precipitate is dissolved in 1.5-2.0 N hydrochloride and excess potassium iodide; a suitable amount of chloroform or benzene is then added. The iodine which is thus liberated and dissolves in the organic layer is titrated with standard sodium thiosulfate. The error is $\leq 0.5\%$ [2].

Method 3. Thallium(III) is titrated potentiometrically or amperometrically with 0.02-0.03 M thiourea. The method is based upon the reduction of thallium(III) to thallium(I) by the thiourea. The equivalence point in an acid or neutral medium is found at a ratio of Tl:thiourea = 1:2. The titration is carried out at pH 1.0-1.4 at 1.0 V (SCE). As little as 1 γ can be estimated amperometrically and 10 γ potentiometrically in 40 ml of solution. Manganese(VII) and chromium (VI) do not interfere [3].

Method 4. Thallium(III) chloride is titrated with standard tin(II) chloride using a Pt-SCE system. Samples containing 3-500 mg/ 25 ml of thallium are titrated with 0.01-0.02 M stannous chloride. Cadmium, copper, arsenic(III), iron(II), iron(III), sulfuric acid, and nitric acid do not interfere. The error in the determination is $\leq$ 1% [4].

Method 5. Samples containing 0.00083-0.53 gm of thallium are titrated with standard tin(II) chloride solutions iodometrically using

a starch indicator. Copper(II), zinc(II), cadmium(II), iron(II),
iron(III), lead, silver, arsenic(III), arsenic(V), antimony(III), anti-
mony(V), hydrochloric acid, and chloride, sulfate, and nitrate ions,
as well as small quantities of sulfuric acid and nitric acid, do not
interfere. The method is rapid, less than 10 minutes per sample, and
is adaptable to automatic titration. The error is less than 1% [5].

Complexometric Titrations

Method 1. Thallium(III) is titrated with standard EDTA using
the iron(III)-sulfosalicylate complex as the indicator. The method is
particularly well suited for both micro- and macro-amounts of thallium.
The determination is carried out at a pH of about 2.6. At the end
point there is a sharp color change from violet to colorless. Calcium,
strontium, barium, magnesium, manganese, mercury(II), chromium(III),
and aluminum do not interfere. Interferences are shown by copper,
nickel, cobalt, bismuth, gallium, indium, and cerium.

Method 2. Thallium(III) and thallium(I) in a mixture of their
chlorides are determined individually. In one aliquot, thallium(III)
is titrated with standard EDTA using xylenol orange as the indicator.
In a second aliquot, thallium(I) is titrated potentiometrically with
standard potassium bromate. In a third aliquot, the sum of the thal-
lium(III) and thallium(I) is determined by titration with standard EDTA
after prior oxidation of the thallium(I) to thallium(III) by evaporation of the sample with aqua regia [7].

References

[1]. V. A. Khadeev and S. N. Poluektova, *Nekotovye Vopr. Khim. Teknol.
i Fiz.-Khim. Analiza Neorgan. Sistem, Akad. Nauk Uz. SSR. Otd.
Khim. Nauk 1963*, 174-183.

[2]. A. A. Verdi-Zade, R. N. Yusubov, and N. A. Verdi-Zade, *Uch. Zap.
Azerb. Gos. Univ. Ser. Khim. Nauk, 1966*(1), 35-37.

[3]. Yu. I. Usatenko and L. M. Kutsenko, *Zavodskaya Lab. 32*(4), 398-
400 (1966).

[4]. H. Basinska and D. Tylzanowska, *Chem. Anal. (Warsaw), 11*(5),
995-998 (1966).

[5]. H. Basinska and D. Tylzanowska, *Chem. Anal. (Warsaw), 12*(1), 27-
31 (1967).

[6]. R. C. Aggarwal and A. K. Srivastava, *Bull. Chim. Soc. Jap., 39*
(10), 2178-2180 (1966).

[7]. R. Pribil, V. Vesely, and K. Kratochvil, *Talanta, 8*, 52-54
(1961).

SCANDIUM

Scandium is determined primarily by complexometric titrations.

Synoptic Survey

Classical Methods

None of importance.

Contemporary Methods

Complexometric Titrations. Scandium(III) chloride is titrated
with standard EDTA using alizarin red S, thoron, or murexide as the
indicator [1].
Scandium in the presence of copper(II) and rare earths is titrated photometrically in acid solution with standard EDTA [2].
Scandium is determined by the addition of excess standard EDTA
followed by back-titration with standard mercuric nitrate [3,4].

Outline of Recommended Contemporary Methods for Scandium

Complexometric Titrations

Method 1. Samples containing 0.02-0.04 mole of scandium chloride
are titrated with standard EDTA. The following indicators are recommended together with the optimum pH for the titration: alizarin red
S (4.5-6.4), thoron (4.5-6.5), or murexide (3.5-7.5). The following
ions do not interfere: lithium, sodium, potassium, calcium, strontium,
barium, magnesium, aluminum, silver, zinc, cadmium, mercury, arsenic
(III), antimony(III), cobalt, nickel, platinum(IV), chloride, bromide,
fluoride, chlorate, bromate, iodate, acetate, sulfite, or sulfate.
The precision and accuracy of the method are excellent [1].

Method 2. Scandium in the presence of copper(II) is titrated
photometrically in an acid solution, optimum pH 2.5-3.0, with standard
EDTA. As much as a 60-fold amount (on a molar basis) of rare earths
may also be present. The absorbance of the copper(II)-EDTA complex
at 745 mμ is used to follow the course of the titration. Calcium,
magnesium, uranium(VI), and even significant amounts of iron(II) along
with other metal ions do not interfere. Bismuth(III), hafnium, thor-
ium(IV), zirconium(IV), fluoride, and sulfate interfere. The precision
and accuracy are excellent. The average error is $\pm$ 0.5% [2].

Method 3. Samples containing 0.045-7.0 mg of scandium are analyzed by the addition of excess standard EDTA and by back-titration
in a basic solution at pH 9.5 with standard mercuric nitrate. In
mixtures containing both scandium and mercury(II), the latter is titrated potentiometrically with standard potassium iodide using a silver
amalgam indicator electrode. The sum of the scandium and mercury(II)
is found complexometrically by the addition of excess EDTA and back-

titration with mercuric nitrate. A ternary mixture of scandium, copper, and mercury is analyzed in a similar fashion. The copper is determined iodometrically; in a second aliquot the mercury is determined potentio-metrically; and in a third aliquot the total concentration of the three elements is determined by an EDTA titration using as above an excess of standard EDTA followed by back-titration at pH 9.5 with standard mercuric nitrate solution [3,4].

References

[1]. S. P. Sangal, *Mikrochem. J.*, *9*(1), 38-41 (1965).

[2]. J. S. Fritz and D. J. Pietrzyk, *Anal. Chem.*, *31*, 1157-1159 (1959).

[3]. H. Khalifa and A. Soliman, *Z. Anal. Chem.*, *169*, 109 (1959).

[4]. H. Khalifa and M. M. Khata, *Z. Anal. Chem.*, *191*, 1339-1345 (1962).

YTTRIUM

Methods for the determination of yttrium include complexometric and precipitation titrations.

Synoptic Survey

Classical Methods

None of importance.

Contemporary Methods

Complexometric Titration. Yttrium is titrated with standard EDTA and xylenol orange as the indicator [1].

Precipitation Titrations. Yttrium is titrated amperometrically with cupferron at pH 3.5-6.0 [2].
Yttrium is precipitated as yttrium arsenate. The latter is determined iodometrically [3,4,5].

Outline of Recommended Contemporary Methods for Yttrium

Complexometric Titration

Yttrium samples are titrated with standard EDTA in a solution buffered with urotropine (hexamethylenetetramine) at pH 5.0-8.0 using xylenol orange as the indicator. In samples containing aluminum, the aluminum is masked with sulfosalicylic acid prior to the titration of the yttrium. In such samples, the sum of yttrium and aluminum is found by adding excess standard EDTA, buffering the solution with an acetate buffer at pH 5.0-5.8, and back titrating the excess EDTA with standard zinc sulfate solution in the presence of dithizone as the indicator. The aluminum is found by difference. The analysis of garnet has been carried out by this procedure. The standard deviations are 0.16% for yttrium and 0.02% for aluminum [1].

Precipitation Titrations

Method 1. Samples containing 40 γ/25 ml are titrated amperometrically with standard cupferron solution at pH 3.5-6.0. A rotating microanode and a stationary microcathode are used; the potential difference is 1.2 V; 0.03 M sodium chloride is used as a coagulant. The relative error is $\leq$ 1.5% [2].

Method 2. Yttrium alone or in the presence of scandium or aluminum and over a wide range of concentration is precipitated in 0.1 N hydrochloric acid as the arsenate. The precipitation may also be carried out in 0.1 N nitric acid or 0.2-0.3 N acetic acid. A 5-10-fold excess of Na_3AsO_4 is added to a hot solution of the sample to effect precipitation. The precipitated yttrium arsenate is titrated iodometrically. In the presence of aluminum the average error is 0.1-0.2%. The time required for a determination is between 40 and 50 minutes [3,4,5].

References

[1]. W. Grosskreutz, D. Schultze, and K. T. Wilke, *Z. Anal. Chem.* *232*(4), 278-279 (1967).

[2]. V. D. Vasilenko, *Ukrain. Khim. Zhur.*, *26*, 767-770 (1960).

[3]. G. B. Shakhtakhtinski, G. A. Aslanov, and B. S. Veliev, *Azerb. Khim. Zh.*, *1964*(5), 97-102.

[4]. G. B. Shakhtakhtinski, B. S. Veliev, and G. A. Aslanov, *Azerb. Khim. Zh.*, *1964*(6), 85-90.

[5]. G. B. Shakhtakhtinski, B. S. Veliev, and G. A. Aslanov, *Dokl. Akad. Nauk Azerb. SSR.*, *22*(1), 12-15 (1966).

LANTHANUM

Methods for the determination of lanthanum include complexometric and precipitation titrations.

Synoptic Survey

Classical Methods

None of importance.

Contemporary Methods

Complexometric Titration. Lanthanum is titrated with standard N-benzoylphenylhydroxylamine [1].

Precipitation Titrations. Lanthanum is titrated amperometrically with standard sodium molybdate [2].

Lanthanum is titrated conductometrically with standard selenite [3] or standard sodium tungstate [4].

Lanthanum is titrated potentiometrically with standard sodium tungstate [5], or sodium orthovanadate [6].

Outline of Recommended Contemporary Methods for Lanthanum

Complexometric Titration

Lanthanum is determined by complexing it with an excess of standard N-benzoylphenylhydroxylamine at pH 8.0, and back-titrating the excess with standard copper sulfate. Each mole of Cu(II) reacts with two moles of the N-benzoylphenylhydroxylamine [1].

Precipitation Titrations

Method 1. Lanthanum nitrate (minimum concentrations 1×10^{-3} M) is determined by direct or reverse amperometric titration with standard sodium molybdate at pH 5-6 in aqueous or aqueous alcohol solutions at an applied potential of -1.5 V versus saturated calomel electrode. The sharp peak at the end point corresponds to the formation and precipitation of the normal lanthanum molybdate, $La_2O_3 \cdot 3MoO_3$. Cations which form precipitates with molybdate ions such as silver, cadmium, cerium (III), thorium, mercury(I), and mercury(II), as well as those anions which react with lanthanum such as vanadate, tungstate, and chromate, must be absent. The method is simple and rapid. The precision and accuracy are very good [2].

Method 2. Samples containing low concentrations (0.4053×10^{-4} M) are titrated conductometrically with standard selenite in 30% ethyl alcohol. The sharp end point reflects the formation of lanthanum selenite.

$$2 \ La^{+3} \ + \ 3 \ H_2SeO_3 \ \rightarrow \ La_2(SeO_3)_{3(s)} \ + \ 6 \ H^+$$

Those cations which form precipitates with selenious acid such as cerium, thorium, cobalt, nickel, manganese, zirconium, and samarium, as well as those anions which react with lanthanum such as tellurite, vanadate, chromate, tungstate, and molybdate, must be absent. This is probably one of the best modern methods for determination of lanthanum. Precision and accuracy are excellent [3].

Method 3. Lanthanum as well as praseodymium and neodymium are titrated conductometrically with standard sodium tungstate in aqueous ethanol medium. End points are sharp with the formation of the normal tungstate.

$$2 \ La^{+3} \ + \ 3 \ WO_4^{-2} \ \rightarrow \ La_2(WO_4)_3 \ (s)$$

Cations which precipitate with tungstate, and anions which react with lanthanum must be absent. Since accuracy and precision are good even at low concentrations, the procedure is well-suited to applications on a microscale [4].

Method 4. Lanthanum nitrate as well as cerium, praseodymium, neodymium, and samarium are titrated potentiometrically with sodium tungstate using a glass electrode in conjunction with a standard calomel electrode as the indicator electrode. The titration is carried out at a pH above 6.8. Both direct and reverse titrations may be carried out. This method as well as other analytical methods for the rare earths are becoming more important with the increasing availability of pure individual rare earth samples. The method is simple and accurate even at low concentrations. For samarium the reported error is 0.2-0.4% [5].

Method 5. Lanthanum, cerium, praseodymium, neodymium, and samarium are determined individually by potentiometric titration with standard sodium orthovanadate. The reaction is followed potentiometrically using a glass electrode versus a standard calomel electrode for determination of the end point. A sharp change in emf is noted at the equivalence point corresponding to the precipitation of $LaVO_4$. The titrations are carried out at a pH above 7. The method is simple, rapid, and accurate even at low concentrations. The addition of ethanol to the extent of 50% of the volume of the vanadate solution reduces the solubility of the precipitated rare earth vanadate and affords a sharper end point [6].

References

[1]. V. D. Vasilenko and G. S. Mazharvoskaya, *Ukr. Khim. Zh.*, *31*(1), 101-104 (1965).

[2]. R. S. Saxena and M. Mittas, *Talanta*, *11*(3), 642-652 (1964).

[3]. S. Prasad and V. N. Garg, *J. Electroanal. Chem.*, *11*(1), 72-74 (1966).

[4]. S. Prasad and R. Swarap, *Mikrochim. Ichnoanal. Acta, 1965* (5-6), 1105-1109.

[5]. G. D. Shivahare and R. C. Verma, *Z. Anal. Chem.*, *227*(4), 279-280 (1967).

[6]. G. D. Shivahare and N. D. Joshi, *Z. Anal. Chem.*, *234*(2), 114 (1968).

CERIUM(III)

Methods for the determination of cerium(III) include redox and precipitation titration.

Synoptic Survey

Classical Methods

Redox Titration. Cerium(III) is oxidized to cerium(IV) by various oxidizing agents such as ammonium persulfate or sodium bismuthate. The resulting cerium(IV) ion is then titrated potentiometrically with standard ferrous sulfate [1,2].

Contemporary Methods

Redox Titrations. Cerium(III) is titrated photometrically with standard potassium dichromate in strong phosphoric acid solutions [3].
Cerium(III) is determined by titration with $K_4Fe(CN)_6$ using as the indicator 3,3'-dimethylnaphthidine [4]. With modifications, the method may also be used to analyze mixtures of cerium(III) and lanthanum [5].

Precipitation Titrations. Cerium(III) is titrated amperometrically with standard sodium molybdate solutions [6].
Cerium(III) is titrated conductometrically with potassium tellurite [7] or with standard sodium metavanadate [8].
Cerium(III) is titrated potentiometrically with selenite [9].

Outline of Recommended Contemporary Methods for Cerium(III)

Redox Titrations

Method 1. Cerium(III) is titrated photometrically with standard potassium dichromate in 10.5 M phosphoric acid at 450 mμ or with a blue filter. In samples containing iron(II), the titration gives the sum of the iron and the cerium. If in another aliquot, the iron(II) is determined independently by titration with standard potassium dichromate in 0.5 M sulfuric acid using barium diphenylamine sulfonate as the indicator, then the cerium content is obtained by simple subtraction. The following ions are oxidized by dichromate and must, therefore, be absent: iron(II), vanadium(IV), manganese(II), arsenic(III), antimony(III), uranium(IV), and molybdenum(V). Fluoride does not interfere.
Although cobalt(II), nickel(II), molybdenum(VI), uranium(VI), and cerium(IV) absorb in the same region as chromium(VI), they do not interfere since their absorption remains constant throughout the titration. Thallium(I), chloride, fluoride, sulfuric acid, iron(III),

copper(II), zinc(II), tungsten(VI), and thorium(IV), do not interfere. Nitrate ion slowly oxidizes cerium(III) in strong phosphoric acid solutions and, thus, interferes. Relative errors are 0.4% for iron and 1.0% for cerium. The method is of value in mineral analysis [3].

Method 2. Cerium in samples containing 0.03-0.6 gm are titrated in aqueous ethanol solutions at pH 5-6 (acetate buffer) with $K_4Fe(CN)_6$ using as a redox indicator 3,3'-dimethylnaphthidine. The error in the determination is less than 1%. Magnesium, calcium, chromium(III), chloride, nitrate, and sulfite do not interfere. Iron(III), copper(II), zirconium(IV), and thallium(III) interfere and are removed by extraction from a strongly acidic solution after the addition of cupferron. Zinc also interferes and is removed before titration of the cerium by precipitation with ammonia solutions. All anions which form complexes with cerium interfere in the determination [4].

Method 3. Samples containing both cerium(III) and lanthanum(III) may also be analyzed by a slight modification of the procedure described in the previous paragraph, Method 2. In this case, the sum of the cerium and lanthanum is found by titration with $K_4Fe(CN)_6$ at pH 6 in 25% aqueous ethanol using as the indicator 3,3'-dimethylnaphthidine and an acetate buffer. The cerium(III) and lanthanum(III) precipitate as $KCeFe(CN)_6$ and $KLaFe(CN)_6$, respectively. In a second sample, the cerium(III) in an alkaline solution at 0° is oxidized with $K_3Fe(CN)_6$ and the Ce(IV) hydroxide and lanthanum hydroxide are removed from the solution by filtration. The $Fe(CN)_6^{-4}$ contained in the $KCeFe(CN)_6$ and $KLaFe(CN)_6$ precipitate from the first sample and that formed by oxidation in the second sample is determined by titration with standard potassium permanganate. The latter titration yields the cerium (III) content of the sample, and subtraction from the results of the permanganate titration of the $KCeFe(CN)_6$ and $KLaFe(CN)_6$ precipitates yields the lanthanum content of the sample [5]. The relative error is $\leq$ 1%. Cerium(III) and lanthanum solutions between 0.05 and 0.2 M are readily analyzed [5].

Precipitation Titrations

Method 1. Cerium(III) is titrated amperometrically with standard sodium molybdate solutions at pH 5.8-6.2 at an applied EMF of -1.95 V (versus SCE). While conductometric, potentiometric, and pH titrations yield accurate results only with concentrated solutions of reactants, amperometric titrations give very accurate end points over a wide range of concentrations even as low as 10^{-3} M cerous salt. The method is simple, rapid, and accurate. The precipitate formed is yellow, amorphous molybdate, $Ce_2(MoO_4)_3$ [6].

Method 2. Cerium(III) is titrated conductometrically at 30 $\pm$ 0.5° with standard solutions of potassium tellurite.

$$2 Ce(NO_3)_3 + 3 K_2TeO_3 \rightarrow Ce_2(TeO_3)_3 + 6 KNO_3$$

The solution must contain 30% ethanol for accurate results [7].

Method 3. Samples containing 5-15 mg of cerium are titrated

conductometrically at 30 $\pm$ 0.5° with standard sodium metavanadate in 15% ethanol. Apparently Ce(NO$_3$)$_3$ is formed. The addition of ethanol is required because in purely aqueous solution the sharpness of the break of the titration curve is no longer observed due to the dissolution of the precipitate and its hydrolysis. The method is simple and precise for the determination of small amounts of cerium(III) [8].

Method 4. Cerium(III) in concentrations of 4 x 10^{-4} M to 2 x 10^{-3} M are titrated potentiometrically against 0.1314 M H$_2$SeO$_3$ in 70% ethanol. Breaks in the titration curve in both direct and reverse titrations correspond to a ratio of 2 moles Ce : 3 moles Se. This corresponds to the formation of Ce$_2$(SeO$_3$)$_3$ [9].

References

[1]. H. H. Willard and P. Young, *J. Am. Chem. Soc.*, *50*, 1397 (1928).

[2]. N. H. Furman, *J. Am. Chem. Soc.*, *50*, 755 (1928).

[3]. G. G. Rao and P. K. Rao, *Talanta*, *14*(1), 33-43 (1967).

[4]. H. Basinska and J. Soboczynska, *Chem. Anal. (Warsaw)*, *11*(5), 989-994 (1966).

[5]. H. Basinska and J. Soboczynska, *Chem. Anal. (Warsaw)*, *12*(1), 1165-1169 (1967).

[6]. R. S. Saxena and M. L. Mittal, *J. Electroanal. Chem.*, *5*, 287-289 (1963).

[7]. S. Prasad and S. T. Kumar, *Proc. Inst. Chemists (India)* *35*, 4-9 (1963).

[8]. S. Prasad and R. C. Srivastava, *Z. Anal. Chem.*, *206*(3), 171-174 (1964).

[9]. S. Prasad and S. Kumur, *J. Indian Chem. Soc.*, *40*(6), 445-450 (1963).

CERIUM(IV)

Methods for the determination of cerium(IV) include redox, complexometric, and precipitation titration.

Synoptic Survey

Classical Methods

None of importance.

Contemporary Methods

Redox Titrations. Cerium(IV) is titrated potentiometrically with standard solutions of molybdenum(III) chloride [1], or chromium(II) chloride [2].

Cerium(IV) and iron(III) in a mixture are titrated amperometrically in succession with standard ascorbic acid [3,4].

Cerium(IV) and vanadium(V) or molybdenum(VI) in a mixture are titrated amperometrically in succession with standard ascorbic acid [5].

Complexometric Titration. Cerium(IV) is titrated potentiometrically with standard thioglycolic acid [6].

Precipitation Titration. Cerium(IV) is precipitated as the arsenate which in turn is determined iodometrically [7].

Outline of Recommended Contemporary Methods for Cerium(IV)

Redox Titrations

Method 1. Cerium(IV) is titrated potentiometrically with standard molybdenum(III) chloride in 2 N sulfuric acid solution or 9 N hydrochloric acid solution. The $MoCl_3$ is best prepared by electrolytic or mercury reduction of the Mo^{+6} ion. Since molybdenum blue is formed at the end point, the titration with practice may be carried out visually. The error is less than 1% [1].

Method 2. Cerium(IV) is titrated potentiometrically with standard chromium(II) solutions in 10 N sulfuric acid with platinum as the indicator electrode versus a standard calomel electrode. The sample must be free of oxygen and is, therefore, purged with carbon dioxide for 15 to 20 minutes prior to titration. The following ions do not interfere: ammonium, potassium, sodium, calcium, barium, silver, aluminum, manganese, cadmium, zinc, nickel, cobalt, europium, yttrium, erbium, terbium, and samarium. Nitrate and most organic acids except formic and acetic interfere [3].

Method 3. Solutions of cerium(IV) in 2.5 M sulfuric acid are titrated with standard ascorbic acid at 50° using a rotating platinum electrode at 0.36 V versus SCE. If iron(III) is present in the sample, the cerium(IV) is first titrated, the pH is then adjusted with ammonia to pH 1.5 and the iron is titrated at 50° with ascorbic acid [3].

Method 4. Cerium(IV) when present in a mixture containing iron(III) is titrated amperometrically in 2.5 M sulfuric acid at 50° using two polarized electrodes at 200 mV. After the titration of the cerium(IV) has been finished, the pH is lowered to 1.5 and the iron (III) is titrated at 100 mV. Results are fairly accurate and precise. The average error is ± 1.0% [4].

Method 5. Mixtures of cerium(IV) and vanadium(V), or mixtures

of cerium(IV) and molybdenum(VI) are titrated amperometrically with
standard ascorbic acid in sulfuric acid. Then, depending on the sample, the vanadium is titrated at pH 1.0 at 100 mV or the molybdenum is
titrated at pH 5.6-6.2 also at 100 mV [5].

Complexometric Titration

    Cerium(IV) in concentrations of 0.0001 - 1 M but preferably in
a 0.02 M solution is titrated potentiometrically with the platinum
calomel electrode and standard thioglycollic acid at pH 1-2. Since
iron(III), copper, cadmium, nickel, lead, and magnesium do not interfere, the method is ideally suited for determination of cerium in cer-
ium-iron and magnesium-cerium alloys [6].

Precipitation Titration

    Cerium(IV) is precipitated as ceric arsenate, $Ce_3(AsO_4)_4$ by the
addition of sodium arsenate. The precipitate is removed and dissolved
in 2:5 v/v sulfuric acid : water. After the addition of potassium
iodide and benzene, the arsenate is titrated with standard sodium thiosulfate [7].

References

[1]. M. Y. Farah and S. Z. Mikhail, *Z. Anal. Chem.*, *166*, 24-31 (1959).

[2]. E. L. Krukovskaya, Sh. T. Talipov, and L. Chimova, *Nauchu Tr.*,
      Tashkentsk. Gos. Univ., No. *264*, 56-63 (1964).

[3]. D. Singh, A. Varma, and V. S. Agarwala, *Z. Anal. Chem.*, *183*,
      172-177 (1961).

[4]. D. Singh and A. Varma, *Current Sci. (India)*, *30*, 137-138 (1961).

[5]. D. Singh and U. Bhatnagar, *Israel J. Chem.*, *5*(1), 29-32 (1967).

[6]. Y. Takeuchi, *Nippon Kagaku Zasshi*, *82*, 1647-1649 (1961).

[7]. I. A. Mamedov and M. N. Nabiev, *Uch. Zap. Azerb. Gos. Univ. Ser.*
      Khim. Nauk, No. *3*, 10-13 (1966).

OTHER LANTHANIDE RARE EARTHS

    Due to the great similarities in atomic structure, the analytical
reactions of the rare earths are often identical. This is also true
for the elements preceding the rare earths in the periodic table,
namely, scandium, yttrium, and lanthanum. Note that many of the listed
titrimetric procedures used to determine the sum of two or more of the
rare earths are also applicable to the titration of an individual element of this group.

The principal methods used in the determination of the rare earths include redox, complexometric, and precipitation titrations.

Synoptic Survey

Classical Methods

Redox Titration. Rare earth nitrates are titrated potentiometrically with standard sodium oxalate. The method is limited to the analysis of the lanthanides or cerium group of rare earths [1].

Contemporary Methods

Complexometric Titrations. Rare earths and thorium are titrated photometrically with standard EDTA using alizarin red S as the indicator.
    Rare earths in the presence of aluminum masked with acetylacetone are titrated with standard ethylenetrinitrilopentaacetic acid, DTPA, and xylenol orange as the indicator [3].
    Rare earths in the presence of iron and thorium are titrated with standard EDTA to a biamperometric end point [4].

Precipitation Titration. Rare earths are titrated with standard sodium tungstate using bromocresol purple as the indicator [5].

Outline of Recommended Contemporary Methods for the Rare Earths

Complexometric Titrations

Method 1. Rare earths and thorium are titrated photometrically with standard EDTA and alizarin red S as the indicator. A light filter with a maximum transmittance at 520 mμ and a 0.5 band width of 50 mμ is employed in end-point detection. The method may be used for concentrations as low as 5×10^{-6} M. In practice the method is limited to the determination of the elements cerium through europium (atomic numbers: 58-63) simultaneously with thorium. Results with gadolinium and terbium are inconclusive.
    Interferences due to cerium are removed by the addition of hydroxylamine. Thorium is titrated first at a pH of 2.8, the pH is then raised to 3.7, and the rare earths are titrated at 85°. With proper variations of the basic method, as little as 0.1% of thorium in a rare earth mixture, or 0.2% of rare earths in thorium can be determined. The method is recommended for the analysis of naturally occurring materials after the usual preliminary separations. Precision and accuracy are good [2].

Method 2. Rare earths are determined conveniently in the presence of large excesses of aluminum by masking the aluminum with acetylacetone and titrated with standard DTPA using xylenol orange as the indicator. As much as a 500:1 M ratio of aluminum to rare earths may be present. The titration mixture is buffered with hexamethylenetetramine at a pH 5-5.5. At the end point the color change is from redpurple to yellow. Precision and accuracy are satisfactory [3].

Method 3. Rare earths can be titrated biamperometrically in the presence of iron and thorium with standard EDTA. Two stationary platinum electrodes are used. Iron and thorium are titrated at pH 1.5-2.0 followed successively by the rare earths of the lanthanum group at pH 5.0. Large amounts of rare earths of the ytterbium group interfere in the iron and thorium determinations. Individual elements are titrated as follows: lanthanum at pH 5.0-8.0, yttrium at pH 3.5-8.0, and ytterbium at pH 3.0-8.0 [4].

Precipitation Titration

Total or individual rare earths are titrated in boiling solution with standard sodium tungstate using either bromocresol purple or a mixture of bromocresol green and methyl red as the indicator. The error is about 0.5% for the mixed indicator and about 1% for bromocresol purple. Heavy metals interfere [5].

References

[1]. G. Jantsch and H. Z. Gawalowski, *Z. Anal. Chem.*, *107*, 389 (1936).

[2]. K. Y. Bril, S. Holzer, and B. Rety, *Anal. Chem.*, *31*(8), 1353-1357 (1959).

[3]. O. I. Milner and S. J. Gedansky, *Anal. Chem.*, *37*(7), 931-933 (1965).

[4]. F. Vydra and F. Horacek, *Anal. Lett.*, *1*(2), 31-37 (1967).

[5]. W. Wawrzyczek and W. Wisniewski, *Z. Anal. Chem.*, *203*, 339-344 (1964).

INDIVIDUAL RARE EARTHS

Synoptic Survey

Classical Methods

None of importance.

Contemporary Methods

Precipitation Titration of Neodymium. Neodymium is titrated with standard sodium tungstate using an absorption indicator [1].

Precipitation Titrations of Samarium. Samarium as the nitrate is titrated conductometrically in 30% ethyl alcohol with standard sodium tungstate [2].

Samarium is titrated conductometrically in 10% alcohol with standard sodium molybdate [3].

Redox Titration of Europium. Europium(III) is determined by controlled potential coulometry [4].

Precipitation Titration of Gadolinium. Gadolinium is titrated conductometrically with standard molybdate [5].

Gadolinium is titrated with standard sodium phosphate using bromocresol purple as the indicator [6].

Redox Titration of Terbium. Terbium is precipitated as the oxalate which yields Tl_4O_7 upon heating. The oxide is determined iodometrically [7].

Complexometric Titration of Dysprosium. Dysprosium is titrated potentiometrically with standard EDTA. Aluminum, if present, is masked with sulfosalicylic acid [8].

Complexometric Titration of Thulium. Thulium is titrated in the presence of α-hydroxybutyric acid and an ammonia buffer at pH 8-9 with standard EDTA [9].

Redox Titration of Ytterbium. Ytterbium is titrated coulometrically in a methanolic solution of tetraethylammonium bromide [10].

Complexometric Titration of Lutetium. Lutetium is titrated with standard EDTA using xylenol orange as the indicator. Iron and aluminum, if present, are masked with ascorbic acid or sulfosalicylic acid respectively [11].

Outline of Recommended Contemporary Methods for Individual Rare Earths

Precipitation Titration of Neodymium

Neodymium in an aqueous solution is titrated at 90° with 0.05 M or 0.1 M sodium tungstate using a mixture of bromocresol green and methyl red as the absorption indicator. The method is fast and precise. The error is much less than 1%. However, the method is of limited value since heavy metals and other rare earths must be absent [1].

Precipitation Titrations of Samarium

Method 1. Samarium nitrate is titrated conductometrically in 30% alcohol with standard solutions of sodium tungstate. For samples containing up to 30 mg of samarium(III) the accuracy is 0.1% [2].

Method 2. Samarium is titrated conductometrically in 10% alcohol solutions with standard sodium molybdate. Sharp breaks in the curve are not found if the titration is carried out in water solutions. The end point is determined graphically. Precision and accuracy are good [3].

Redox Titration of Europium

Europium(III) is determined by controlled potential coulometry. The fact that the oxidation of europium(II) to europium(III) proceeds with 100% current efficiency forms the basis for this analysis. Two steps are required: (1) the reduction of europium(III) to europium(II) at -0.8 V, followed by (2) the oxidation of europium(II) to europium (III) at -0.1 V. The quantity of current needed in step 2 is a measure of europium titrated. Samples containing 5 to 10 mg of europium may be determined. The method is rapid and precise. Samples containing 0.92 and 9.146 mg Eu show results having standard deviations of 0.27 and 0.06% respectively [4].

Precipitation Titrations of Gadolinium

Method 1. Samples containing 4.5×10^{-4} moles of gadolinium are determined by conductometric titration with standard molybdate. Cations which are precipitated by molybdate and anions which react with gadolinium must be absent. Accuracy and precision are excellent. This is probably the most convenient method available at the present time [5].

Method 2. Gadolinium nitrate is titrated with 0.06-0.15 M trisodium phosphate using bromocresol purple as the indicator. This method may also be used for the determination of total rare earths in a sample. The reported error is less than 0.3% [6].

Redox Titration of Terbium

Terbium alone or in the presence of rare earths of the yttrium subgroup is determined iodometrically. Terbium together with yttrium is precipitated as the oxalate. The mixture of oxalates is ashed at 450°, followed by ignition at 500° for five to six hours. The terbium oxide, Tb_4O_7, which is formed is dissolved in a KI solution under an atmosphere of carbon dioxide, the solution is acidified with hydrochloric acid, and the liberated iodine is titrated with standard sodium thiosulfate. The terbium content of the mixtures may run as high as 20%. Accuracy is fair [7].

Complexometric Titration of Dysprosium

Samples containing 3-10 mg of dysprosium are titrated with standard EDTA. Samples may contain up to 1000 mg of aluminum provided that the aluminum is masked prior to titration by the addition of sulfosalicylic acid. The titration is carried out at pH 6.0 using xylenol orange as the indicator. The method is reasonably accurate, and is useful in checking the homogeneity of dysprosium-aluminum alloys [8].

Complexometric Titration of Thulium

Samples containing 1×10^{-5} moles of thulium in 50 ml of solution

to which has been added 1 x 10^{-3} mole of α-hydroxybutyric acid are titrated visually in an ammonia buffer at pH 8-9 with 0.01 M EDTA. The
color change at the end point is from purple to sky-blue. The reported
average deviation is less than 2% [9].

Redox Titration of Ytterbium

Ytterbium in the presence of europium is titrated coulometrically
in methanol solutions of tetraethylammonium bromide. The procedure
involves the one-electron reduction of the respective trivalent ions.
The reduction of ytterbium is induced by the primary reduction of the
europium, so that the sum of both metals is determined in samples containing both. It is then necessary to determine the europium separately by coulometric oxidation in 0.1 N hydrochloric acid. Ytterbium
is then found by difference. The mean absolute error is less than
0.05 µeq in the sample range of 0.5 to 20 µeq [10].

Complexometric Titration of Lutetium

Samples containing about 2 mg of lutetium in the presence of
aluminum are titrated at pH 5.5 $\pm$ 0.1 using xylenol orange as the indicator. The titration is carried out with standard EDTA; the pH of
the solution is adjusted by means of an ammonium buffer. Alloys containing lutetium and aluminum are dissolved in hydrochloric acid and
the aluminum is masked with 20% (w/v) sulfosalicylic acid. For each
milligram of aluminum present, 0.5 ml of the sulfosalicylic acid solution is used. Iron present is masked with ascorbic acid. Color
change at the end point is from red to yellow. The relative standard
deviations are 0.5 and 1.1 for samples containing 4.97% and 0.98%
lutetium respectively [11].

References

[1]. W. Wawrzyczek and A. Bukowska, *Anal. Chim. Acta, 30*(4), 401-403
      (1964).

[2]. S. Prasad and R. Swarup, *Z. Anal. Chem., 209*(3), 397-398 (1965).

[3]. K. P. Srivastava, *J. Proc. Inst. Chemists (India), 37*(1), 9-13
      (1965).

[4]. W. D. Shults, *Anal. Chem., 31*(6), 1095-1098 (1959).

[5]. K. P. Srivastava, *J. Indian Chem. Soc., 44*(1), 42-46 (1967).

[6]. W. Wawrzyczek and A. Benedict, *Chem. Anal. (Warsaw), 9*(3), 621-
      622 (1964).

[7]. G. A. Bornyi, *Ukr. Khim. Zh. 28*, 393-395 (1962).

[8]. A. Brueck and K. F. Lauer, *Anal. Chim. Acta, 39*, 135-136 (1967).

[9].   G. L. Silver and R. C. Bouman, *Talanta, 14,* 893-896 (1967).

[10]. E. N. Wise and E. J. Cokal, *Anal. Chem., 32*(11), 1417-1419
       (1960).

[11]. A. Brueck and K. F. Lauer, *Anal. Chim. Acta, 33,* 338-340 (1965).

THORIUM

Methods for the determination of thorium include an indirect
redox and various complexometric titrations.

Synoptic Survey

Classical Methods

Redox Titration. Thorium is precipitated as the molybdate which
is reduced with amalgamated zinc. The molybdenum(III) so formed is
added to excess ferric alum solution and the ferrous ions formed are
titrated with standard ceric sulfate solution [1].

Contemporary Methods

Complexometric Titrations. Thorium is titrated with standard
1-hydroxyethylidene-1,1-diphosphonic acid, HEDPHA, in the presence
of 1,2-diaminocyclohexanetetraacetic acid, DCTA. The titration is
carried out in a solution buffered with urotropine using xylenol
orange as the indicator [2].
      Thorium is titrated spectrophotometrically with standard EDTA
and using 3-(6-sulfo-2-naphthyl-azo)-chromotropic acid, β-SNADNS-6,
as the indicator [3].
      Thorium in the presence of the rare earths is titrated at pH 2
with standard triethylenetetraamine hexaacetic acid, TTHA, and xylenol
orange or TMS as the indicator [4].
      Thorium in the presence of scandium is determined by titrating
the sum of the two elements with standard EDTA and xylenol orange as
the indicator. Excess triethylenetetraamine hexaacetic acid, TTHA,
is then added and the EDTA displaced from the scandium complex plus
the excess TTHA are titrated with standard zinc solution [5].
      Thorium in the presence of scandium is determined by titrating
two aliquots. In the first, the sum of the two elements is determined
by titration with standard diaminocyclohexanetetraacetic acid, DCTA.
To the second aliquot, an excess of triethylenetetraminehexaacetic
acid, TTHA, is then added, and in the presence of disodium hydrogen
phosphate, back-titration with standard zinc solution gives values
for excess TTHA and the scandium displaced from its TTHA complex [6].

Outline of Recommended Contemporary Methods for Thorium

Complexometric Titrations

Method 1. Thorium is titrated with 0.025 M 1-hydroxyethylidene-
1,1-diphosphonic acid, HEDPHA, in the presence of DCTA. The reaction
involved is the formation of the soluble binuclear ternary complex,
$Th_2(DCTA)_2(HEDPHA)$. The titration is carried out in a weakly acidic
solution buffered with urotropine and using xylenol orange as the indicator. The DCTA is a perfect masking agent for trivalent metals, e.g.,
scandium, yttrium, gallium, indium, bismuth, and iron. The following
elements interfere: aluminum, zirconium, titanium, and large amounts
of thallium(III) [2].

Method 2. Samples of thorium containing 10-130 µg in 5 ml are
titrated spectrophotometrically in a solution buffered with glycine-
HCl at pH 2.5-3. Titration is carried out with standard EDTA and β-
SNADNS-6 as the indicator at a wave length of 590 mµ. Equal amounts
by weight of the following metals do not interfere: silver, copper(II),
lead, mercury(II), zinc, and iron(II), alkali metals and alkaline
earths may be present up to fourfold the amount of thorium without
interfering. Cobalt(II), nickel, manganese(II), and uranium(VI) interfere only slightly when present at one half the weight of the thorium.
Traces of beryllium, aluminum, zirconium, iron(III), fluoride, sulfate,
phosphate, oxalate, citrate, and nitrate interfere. Iron(III) may be
masked by the addition of ascorbic acid or hydroquinone. The method
is reasonably fast. Precision and accuracy are excellent [3].

Method 3. Thorium when present in a mixture with the rare earths
is determined independently by titration at pH 2 with standard TTHA
using xylenol orange or 3',3" bis-{[bis-(carboxymethyl)amine]-methyl}
thymolsulfonaphthalein, TMS, as the indicator. To determine the rare
earths, the pH is then increased to 6 and an excess of standard TTHA
is added. The excess is back-titrated with standard zinc chloride
solution [4].

Method 4. Thorium and scandium in the presence of each other
may be determined by a procedure which makes use of the fact that the
scandium-EDTA complex is not affected by the addition of excess TTHA
whereas the thorium-EDTA complex liberates EDTA forming the thorium-
TTHA complex. The procedure involves titration with standard EDTA at
pH 2.5-3.5 using xylenol orange as the indicator to determine the sum
of the two metals. An excess of TTHA is then added which displaces
the EDTA from the thorium-EDTA complex. Finally, the liberated EDTA
and excess TTHA are titrated with standard zinc solution. From this
data, the thorium and scandium values are calculated [5].

Method 5. A variation on method 4 above involves the titration
of both thorium and scandium with 0.05 M DCTA at pH 2.5-3 and at a
temperature of 50° using xylenol orange as the indicator. To a separate aliquot of the mixture, there is added an excess of TTHA (0.5 M)
enough 20% urotropine to adjust the pH to 5-5.5, a few drops of xylenol orange, and 1 M disodium hydrogen phosphate. The excess TTHA and
the scandium displaced from its TTHA complex is then titrated with
0.05 M zinc solution. At the end point the color change is from yellow
to an intense red-violet [6].

References

[1]. C. V. Banks and H. Diehl, *Anal. Chem.*, *19*(4), 222 (1947).

[2]. R. Přibil and V. Veselý, *Talanta*, *14*, 591-595 (1967).

[3]. S. K. Datta and S. N. Saha, *Chemist-Analyst*, *51*, 49-50 (1962).

[4]. A. K. Mukherji, *Talanta*, *13*(8), 1183-1185 (1966).

[5]. R. Přibil and V. Veselý, *Talanta*, *11*, 1545-1547 (1964).

[6]. R. Přibil, V. Veselý, and J. Horácek, *Talanta*, *14*(2), 266-268 (1967).

URANIUM

Methods for the determination of uranium include redox, complexo-metric, and precipitation titrations.

Synoptic Survey

Classical Methods

Redox Titrations. Uranium(VI) is reduced to uranium(IV) with zinc or bismuth amalgam, followed by titration with standard potassium permanganate [1].
Uranium(VI) is reduced to uranium(IV) using a lead reductor. The uranium(IV) is determined indirectly by adding excess ferric ion and titrating the ferrous ion formed with standard potassium dichromate [2].

Precipitation Titration. Uranium(VI) is titrated with standard disodium hydrogen phosphate [3].

Contemporary Methods

Redox Titrations. Uranium(VI) is reduced with chromium(II). Excess ferric ion is added and the equivalent amount of ferrous ion formed is titrated potentiometrically with standard ceric sulfate solution [4].
Uranium(VI) is reduced on the Jones reductor to a mixture of uranium(III) and uranium(IV). These ions are determined by controlled current bipotentiometric titration or amperometrically with standard iron(III) solution [5,6].
Uranium(VI) is reduced coulometrically to uranium(IV) in neutral or slightly acidic sodium fluoride solution.
Finally, the uranium(IV) is determined by controlled-potentio-metric coulometric titration [7].
Uranium(IV) and uranium(VI) are determined in a mixture by

titration of the uranium(IV) in phosphoric acid solution with potassium
dichromate. Excess titanium(III) is then added to reduce the uranium
(VI) to uranium(IV). The determination is concluded by titration of
the uranium(IV) thus formed potentiometrically at 95° with standard
potassium dichromate [8].

Complexometric Titration. Uranium(VI) is titrated photometrically with standard EDTA using PAN indicator at 555 mμ, or without the
indicator in the UV region [9].

Outline of Recommended Contemporary Methods for Uranium

Redox Titrations

Method 1. Uranium(VI) is reduced with chromium(II) in sulfuric
acid solution. The excess reducing agent is then removed by an air
purge. Excess ferric ions are added, and the equivalent amount of
ferrous ions formed is titrated potentiometrically with standard ceric
sulfate solution. Automatic titration is feasible. If iron is present, the above technique gives the total of the iron and uranium
present. Iron must then be determined separately and the value deducted
from the total. The relative precision at the 50 mg level of uranium
is better than ± 0.5% [4].

Method 2. Uranium(VI) is reduced in a Jones reductor to a mixture of uranium(III) and uranium(IV). The mixture is determined by
bipotentiometric titration with standard iron(III) sulfate solution.
Two platinum electrodes of 26-gauge wire 1-cm long are used as indicator electrodes. A pH-meter serves both as the source of the 10 μA
polarizing current applied to the electrodes and as a voltmeter to
measure the potential between the electrodes as a function of added
titrant. The method is reasonably fast requiring about 45 minutes for
one determination. The standard deviation for samples containing
26.17 mg uranium is 0.1% [5].

Method 3. Uranium(VI) is reduced to a mixture of uranium(III)
and uranium(IV) with zinc. The mixture is then titrated amperometrically using a rotating platinum electrode at −50 mV versus SEC in 0.3
N sulfuric acid with standard iron(III) under an inert atmosphere.
The titration is carried out first to the uranium(III)-(IV) end point
and then to the uranium(IV)-(VI) end point. Prior to the first end
point, the current is anodic, between the end points it is less than
1 μA, and after the second end point it is cathodic. Vanadium and
molybdenum interfere. Iron does not interfere when present in iron
to uranium weight ratios of as high as 8 to 1. Chromium in weight
ratios of chromium to uranium of 1 or less does not interfere provided
that the chromium is oxidized to chromium(VI). The relative standard
deviation is 0.1% for 200 mg of uranium and 0.3% for samples containing 20 mg uranium [6].

Method 4. Uranium is determined by controlled-potential coulometric titration of uranium(VI) at −1.000 V versus SCE in neutral or
very weakly acidic, pH 6.5-7.0, solutions of 0.75 M sodium fluoride.

Chromium(III), copper(II), molybdenum(VI), and zirconium in moderate amounts do not interfere, nor does iron(III) in a molar ratio of 1:1. Aluminum shows no interference in weight ratios of aluminum to uranium of 5 to 1. The relative error is +0.2-0.3% at the 6 mg uranium level; the relative standard deviation is $\leq$ 0.5% [7].

Method 5. In a mixture of uranium(IV) and uranium(VI) formed from the oxides and dissolved in concentrated phosphoric acid, uranium (IV) is titrated potentiometrically with 0.03 N potassium dichromate. An excess of titanium(III) is then added to reduce the uranium(VI) to uranium(IV). The solution is then heated to 95°, and again titrated potentiometrically with the same dichromate solution [8].

Complexometric Titration

Uranium as UO_2^{+2} is titrated with 0.033 M EDTA at pH 4.4 using PAN as the indicator. The end point is determined spectrophotometrically at 555 mμ. The titration can also be carried out in the absence of the indicator in the ultraviolet region at 320 mμ. The method is very precise in the range of 12 μg - 9 mg uranium. Because of the rather low dissociation constant of the uranium-PAN complex, the EDTA solution must be standardized with a standard uranium sample at a concentration close to that being analyzed. For the concentration range of 12 μg to 9 mg of uranium, the coefficients of variation are between 5 and 0.1% [9].

References

[1]. K. Someya, *Science Repts., Tohoku Imp. Univ.*, *15*(1), 411 (1926).

[2]. W. D. Cook, F. Hazel, and W. M. McNabb, *Anal. Chem.*, *22*, 654 (1950).

[3]. N. H. Furman, ed., *Standard Methods of Chemical Analysis*, 6th Ed., Vol. 1, D. Van Nostrand, Princeton, N. J., 1962, p. 1194.

[4]. S. L. Jones, *U.S. At. Energy Comm.*, *KAPL-M-SLJ-7, 24pp., (1961)*, see: *Nucl. Sci. Abs. 16*(15), *Abs. No. 18811 (1962)*.

[5]. R. A. Whiteker and D. W. Murphy, *Anal. Chem.*, 39(2), 230-232 (1967).

[6]. R. F. Sympson, R. P. Larsen, R. J. Meyer, and R. D. Oldham, *Anal. Chem.*, *37*(1), 58-60 (1965).

[7]. W. R. Mountcastle, Jr., L. B. Dunlap, and P. F. Thomasen, *Anal. Chem.*, *37*(3), 336-340 (1965).

[8]. F. Nakashima and K. Saskai, *Nippon Kagaku Zasshi*, *85*(1), 40-44 (1964).

[9]. A. Brueck and K. F. Laner, *Anal. Chim. Acta*, *37*(3), 325-331 (1967).

NEPTUNIUM

Contemporary methods for the determination of neptunium include redox and complexometric titrations.

Synoptic Survey

Classical Methods

None of importance.

Contemporary Methods

Redox Titrations. A mixture containing neptunium(IV), neptunium(V), and neptunium(VI) is analyzed by controlled potential coulometry [1].
Neptunium and uranium in the presence of each other are determined by controlled potential coulometry [2].

Complexometric Titration. Neptunium(IV) is titrated with standard EDTA using xylenol orange as the indicator [3].

Outline of Recommended Contemporary Methods for Neptunium

Redox Titrations

Method 1. Since neptunium(IV) is not oxidized and neptunium(V) is not reduced in significant quantities at a platinum electrode, neptunium(IV), neptunium(V), and neptunium(VI) may be determined in a mixture by controlled potential coulometric titration. In the first titration, a potential is applied at the controlled potential electrode to reduce neptunium(VI) to neptunium(V). The second titration involves the electrolytic oxidation of neptunium(V) to neptunium(VI). Excess cerium(IV) is then added to the titration mixture to oxidize neptunium(IV) to neptunium(VI). By means of electrolytic reduction, as the final step, the neptunium(VI) and excess cerium(IV) are reduced to neptunium(V) and cerium(III). The integrated current for the first titration gives the neptunium(VI) value. The difference between the integrated currents for the first and second titrations yields the neptunium(V) value. The neptunium(IV) value is the result of the difference between the integrated currents for the second and final titrations [1].

Method 2. The determination of neptunium and uranium in an uranium-neptunium alloy is based on a controlled potential coulometric titration. The determination of neptunium is carried out in a cell with a platinum-gauge as the working electrode. Samples containing 0.5-5.0 mg neptunium are placed in the titration cell which also

contains cerium(IV) sulfate and sulfamic acid. The neptunium sample
is reduced at 0.66 V versus SCE under argon until the cell current is
decreased to 30 μA. The solution is then oxidized at 1.02 V versus
SCE until the current is decreased to 30 μA. Using the integrated
current consumed during oxidation, the amount of neptunium is calculated
for the one-electron oxidation. The determination of uranium is carried out in a cell using mercury as the working electrode. An aliquot
containing 1.0-6.0 mg uranium is placed in the titration cell together
with 0.05 M cerium(IV) sulfate solution. The solution, after prereduction at 0.085 V versus SCE under argon until cell current is decreased
to 50 μA, is then reduced at -0.325 V versus SCE until the cell current
again decreased to 50 μA. The uranium content is then calculated from
the integrated current consumed during the last reduction. The recovery of the neptunium standard in the presence of uranium is 100-102%
with a relative standard deviation of 0-13%. The recovery of a uranium
standard in the presence of neptunium is 100.03% with a relative standard deviation of 0.13% [2]. Gold, platinum, mercury, thorium, and
chloride interfere at concentrations of 0.1 M or higher with the neptunium titration. Plutonium or palladium cause only a slight interference. The prereduction step eliminates most more electropositive
ions such as iron(III), neptunium(VI), plutonium(IV), or cerium(IV)
which interfere in the uranium determination. The main interfering
ions are those which react at the cathode during the uranium titration,
namely, antimony(III), bismuth(III), copper(I), and molybdenum(VI) [2].

Complexometric Titration

Neptunium(IV) in amounts from 1-4 mg is titrated with 0.112 M
EDTA using xylenol orange as the indicator at pH 1.3-2.0. The complex formed shows a 1:1 ratio. Iron(III) and thorium interfere.
There is no interference from the following ions: magnesium, zinc,
cadmium, lead(II), alkali metals, alkaline and rare earths, nitrate,
acetate, sulfate, oxalate, and dichromate. The following ions may be
present but in amounts less than stated: chromium(III), 20 mg; nickel
(II), 15 mg; cobalt(II), 20 mg; uranyl, 50 mg [3].

References

[1].   R. W. Shomatt, *Anal. Chem.*, *32*, 134 (1960).

[2].   C. E. Plock and W. S. Polkinghorne, *Talanta*, *14*(11), 1356 (1967).

[3].   A. P. Smirnov-Averin, G. S. Kovalenko, N. Ermolaev, and N. N.
       Krot, *Zn. Analit. Khim.*, *21*(1), 76-78 (1966).

PLUTONIUM

Methods for the determination of plutonium include redox and
complexometric titrations.

Synoptic Survey

Classical Methods

Redox Titrations. Plutonium(III) usually formed by reduction
with zinc amalgam or titanium(III) is titrated with standard cerium(IV),
permanganate, or dichromate [1].

Plutonium(VI) is determined by reduction to plutonium(IV) with
electrolytically generated iron(II) [1,2].

Contemporary Methods

Redox Titrations. Plutonium is determined by constant current
potentiometry with standard potassium dichromate [3,4].

Plutonium is oxidized to the hexavalent state using perchloric
acid. Excess standard iron(II) is added and the excess is back-
titrated with standard ceric(IV) solution [5].

Plutonium(VI) is reduced by excess electrolytically generated
iron(II). The excess is back-titrated electrolytically [6].

Complexometric Titration. Plutonium(IV) dissolved in acetone is
treated with excess standard EDTA and back-titrated photometrically
with standard zinc chloride using dithizone as the indicator [7].

Outline of Recommended Contemporary Methods for Plutonium

Redox Titrations

Method 1. Plutonium is determined by constant-current potentiometric titration with standard potassium dichromate. The plutonium
metal is dissolved in 4 N sulfuric acid and reduced to plutonium(III)
in a Jones reductor. Polarized gold electrodes are used to indicate
the end point in the titration with primary standard dichromate.
Direct weight titration may also be used with a constant-current potentiometric determination of the end point. Relative standard deviation is 0.04% or better for samples containing 250-450 mg plutonium.
For smaller samples, the relative standard deviation is about 0.1%.
Cations such as chromium, iron, molybdenum, titanium, uranium, vanadium, and tungsten which are reduced in a Jones reductor interfere.
There is no significant interference from americium [3,4].

Method 2. Samples containing 200 or more milligrams of pluton-
ium(VI) are titrated with a slight excess of standard iron(II) ions.
If present in a lower valence state, the plutonium is first oxidized
to plutonium(VI) by means of perchloric acid. The excess standard
iron(II) is back-titrated automatically with standard ceric solution.
For a high purity sample, an average plutonium content of 99.98% was
found with a standard deviation of 0.02%. Large samples and the use
of weight burets are to be preferred for precise results. From the
common metals usually accompanying plutonium, only chromium, manganese, vanadium, gold, and platinum interfere. The method is well-
suited for the analysis of high purity plutonium metal samples and
alloys with noninterfering metals such as iron. It should be noted

that the end point may also be determined spectrophotometrically at
380 mμ.   Standard deviation is 0.02% [5].

   Method 3. Plutonium is determined indirectly in the presence
of large amounts of iron and uranium. The plutonium(VI) is reduced
with electrolytically generated iron(II). An excess is used followed
by an electrolytic back-titration. The relative standard deviation
for samples containing 25.59 mg plutonium is 0.1% [6].

Complexometric Titration

   Plutonium(IV) in nitric acid solutions of irradiated uranium is
extracted with the quaternary amine Hyamine 1622 (Röhm and Haas) dissolved in benzene. After removal of the benzene by evaporation, the
residue is dissolved in acetone and an excess of standard EDTA is
added. Uncomplexed EDTA is back-titrated with standard zinc chloride
solution using dithizone as the indicator. The color change from green
to pink at the end point is observed photometrically, and found graphically as the point of inflection on plotting the galvanometer reading
against volume of added reagent. The standard deviation is 0.4% for
samples containing 100 γ of plutonium [7].

References

[1]. C. F. Metz, *Anal. Chem., 29*(12), 1748 (1957).

[2]. W. N. Carson, Jr., J. W. Vanderwater, and H. S. Gile, *Anal.
      Chem., 29*, 1417 (1957).

[3]. C. E. Pietri and J. A. Baglio, *Talanta, 6*, 159-160 (1960).

[4]. C. E. Pietri and A. W. Wenzel, *Talanta, 14*, 215-218 (1967).

[5]. G. R. Waterbury and C. F. Metz, *Anal. Chem., 31*(7), 1144-1148
      (1959).

[6]. W. D. Shults, *Anal. Chem., 33*, 15-18 (1961).

[7]. D. G. Boase, J. K. Foreman, and J. L. Drummond, *Talanta, 9*,
      53-63 (1962).

GROUP IV

Carbon	Titanium
Silicon	Zirconium and Hafnium
Germanium	
Tin	
Lead	

CARBON

Titrimetric analysis of only carbon dioxide, cyanide, and cyanate will be discussed. The determination of carbon in these forms depends on acid-base, complexometric, or precipitation titrations.

Synoptic Survey

Classical Methods

Acid-Base Titration. Carbon dioxide is absorbed in barium hydroxide and the precipitated carbonate is titrated with standard hydrochloric acid [1].

Complexometric Titration. Cyanide ion is titrated with silver nitrate in a neutral or basic solution. The sodium silver cyanide formed, $NaAg(CN)_2$, is soluble as long as cyanide is present in excess. Insoluble silver cyanide is formed as soon as there is the slightest excess of silver nitrate [2].

Precipitation Titrations. Cyanide ion is determined by adding an excess of standard silver nitrate. After removal of the precipitated silver cyanide, the excess silver nitrate is titrated with standard thiocyanate by Volhard's method using ferric alum as the indicator [3].
Cyanate is titrated with standard silver nitrate. The end point is detected with an adsorption indicator, or by potentiometric or conductometric techniques [4-7].

Contemporary Methods

Acid-Base Titration. Traces of carbon dioxide are determined by spectrophotometric titration [8].

77

Complexometric Titration. Cyanate is titrated spectrophotometrically in dimethylformamide using standard cobalt(II) solutions [9].

Precipitation Titrations. Cyanide is titrated potentiometrically using bimetallic electrodes [10,11].
Cyanide in the presence of thiocyanate is titrated amperometrically with standard silver nitrate [12].

Outline of Recommended Contemporary Methods for Carbon

Acid-Base Titration of Carbon Dioxide

Trace amounts of carbon dioxide as found in a gas stream are titrated spectrophotometrically. After absorption of the carbon dioxide in an excess of very dilute standard base, the excess of the base is back-titrated with standard but very dilute hydrochloric acid. The titration is carried out using phenolphthalein as the indicator and the end point is monitored at 555 mμ. Sensitivity is about 1 ppm of carbon dioxide for 20 liters of gas mixture. The accuracy is about 0.4 ppm in the 1-10 ppm range [8].

Complexometric Titration of Cyanate

Samples containing 2.5-20.0 mg/ml of cyanate diluted with 50 ml of dimethylformamide are titrated spectrophotometrically with 0.025 M cobalt(II) perchlorate at 642 mμ. The end point is found from a plot of absorbance versus volume of added titrant. For a 10.0 mg sample, the standard deviation is $\pm$ 0.1 mg, and the relative standard deviation is $\pm$ 1% [9].

Precipitation Titrations of Cyanide

Method 1. Free cyanide or in mixtures with argenticyanide is determined by potentiometric titration with 0.02-0.1 N silver nitrate using bimetallic electrodes such as platinum-nickel, platinum-silver, platinum-tungsten, or platinum-palladium. In the titration of mixtures of cyanide and argenticyanide, two characteristic changes in potential occur for each of the four above electrode couples. At the first inflection, there is a noticeable turbidity. This corresponds to the end point for the cyanide titration. The second inflection in the curve indicates the complete precipitation of silver cyanide. Thus, both ions can be determined by the same buret run. For the platinum-tungsten electrode, there is a rapid potential increase at the two end points; for the other three electrodes, there is a decrease in potential and the formation of a reversed S-shaped titration curve [10,11].

Method 2. A mixture of cyanide and thiocyanate is readily analyzed by determining first the sum of cyanide and thiocyanate by amperometric titration in 0.1 N potassium nitrate containing 0.02% gelatin with standard silver nitrate. In another sample, cyanide is masked by adding excess formaldehyde, the excess is oxidized with nitric acid, and the thiocyanate is determined in the usual manner. Error is $\leq$ 2.5%. The method can also be used for the determination of cyanide in halide mixtures [12].

References

[1]. N. H. Furman, ed., *Standard Methods of Chemical Analysis*, 6th Ed., Vol. 1, D. Van Nostrand, Princeton, N. J., 1962, p. 293.

[2]. N. H. Furman, ed., *Standard Methods of Chemical Analysis*, 6th Ed., Vol. 1, D. Van Nostrand, Princeton, N. J., 1962, p. 761.

[3]. N. H. Furman, ed., *Standard Methods of Chemical Analysis*, 6th Ed., Vol. 1, D. Van Nostrand, Princeton, N. J., 1962, p. 761.

[4]. R. Ripan-Tilici, *Z. Anal. Chem.*, *102*, 32 (1935).

[5]. R. Ripan-Tilici, *Z. Anal. Chem.*, *98*, 23 (1934).

[6]. O. Pfundt, *Angew. Chem.*, *46*, 218 (1933).

[7]. R. Ripan-Tilici, *Z. Anal. Chem.*, *99*, 415 (1934).

[8]. J. W. Loveland, R. W. Adams, H. H. King, Jr., F. A. Nowak, and L. J. Cali, *Anal. Chem.*, *31*, 1008-1010 (1959).

[9]. F. Trussell, P. A. Argabright and W. F. McKenzie, *Anal. Chem.*, *39*(8), 1025-1026 (1967).

[10]. K. Ueno and T. Tachikawa, *Bunseki Kagaku*, *7*, 756-761 (1958).

[11]. K. Ueno and T. Tachikawa, *Muroran Kogyo Diagaku Kenyu Mokoku*, *2*, 377-378 (1959).

[12]. S. Musha and S. Ikeda, *Bunseki Kagaku*, *14*(3), 270 (1965).

SILICON

Titrimetric determinations of silicates include acid-base and redox titrations.

Synoptic Survey

Classical Methods

None of importance.

Contemporary Methods

Acid-Base Titrations. Silicate is precipitated as potassium hexafluosilicate which is hydrolyzed to hydrofluoric acid and titrated with standard base [1].

Silicate is precipitated as potassium hexafluosilicate which upon treatment with calcium chloride liberates HCl. The liberated acid is titrated with standard base [2].

Redox Titration. Silicon as soluble silica is converted to $H_4(SiMo_3O_{10})_4$ which, in turn, is reduced. The amount of reducing agent is determined spectrophotometrically [3].

Outline of Recommended Contemporary Methods for Silicon

Acid-Base Titrations

Method 1. Silicon as SiO_2 is fused with potassium carbonate. The resulting cooled melt is dissolved in a 1:1 mixture of hydrochloric and nitric acids, and the K_2SiF_6 is precipitated by the addition of sodium fluoride and an excess of potassium chloride. The precipitate is filtered, washed with 3.5% KCl, and added to an excess of water. The liberated hydrofluoric acid is titrated with 0.06 N sodium hydroxide. Since no evaporation is needed, the method is quite fast and adaptable to automatic titration [1].

Method 2. Silicon is precipitated in the usual manner as potassium hexafluosilicate. The precipitate is added to hot water, treated with calcium chloride, and the liberated hydrochloric acid is titrated with standard base. Tachiro's mixed indicator is used. The time for a determination is between 15 and 20 minutes, and is suitable for the determination of silica in various silicates. Accuracy is ± 0.1% [2].

Redox Titration

Soluble silica is determined by conversion to $H_4(SiMo_3O_{10})_4$ followed by titration with tin(II) ion in sulfuric acid, or iron(II) ion in oxalic acid. The end point in the reduction is found potentiometrically or spectrophotometrically. Results show a 4-electron reduction of the $H_4(SiMo_3O_{10})_4$ so that 1 ml of 0.01 N reducing agent is equal to 0.152 mg SiO_2. The method has been recommended in water analysis, and is reported to be fairly accurate for practical purposes [3].

References

[1]. J. Louvrier and I. A. Voinovitch, *Ind. Céram.*, *510*, 243-247 (1959).

[2]. B. Bieber and Z. Večeřa, *Hutmcké Listy*, *15*, 397-398 (1960).

[3]. T. Takahashi and S. Miyake, *Talanta*, *4*, 1-7 (1960).

GERMANIUM

Methods for the determination of germanium include acid-base, complexometric, and precipitation titrations.

Synoptic Survey

Classical Methods

Acid-Base Titration. Germanium(IV) reacts with mannitol to form a complex monobasic acid which is titrated with standard base using phenolphthalein as the indicator [1].

Contemporary Methods

Acid-Base Titrations. Germanium as the potassium hexafluoro-germanate is hydrolyzed and the liberated hydrofluoric acid is titrated with standard base [2,3].

Germanium is determined by a modified "classic method" using better indicators, or a potentiometric approach [4].

Germanium solutions are treated with excess pyrocatechol, and the excess pyrocatechol not bound in the germanium-complex is titrated with standard base [5].

Complexometric Titration. Germanium is titrated amperometrically with standard pyrocatechol [6].

Precipitation Titration. Germanium(IV) is titrated amperometrically with standard magnesium sulfate [7].

Outline of Recommended Contemporary Methods for Germanium

Acid-Base Titrations

Method 1. Germanium in the sample is precipitated as the potassium hexafluogermanate from a solution consisting of 10% hydrogen fluoride and 2.5 N hydrochloric acid. The hexafluogermanate is hydrolyzed by the addition of excess water, and the hydrofluoric acid so formed is titrated with standard base. Error is 0.02% and the reproducibility is $\pm$ 0.5% for samples containing 20-160 mg germanium. Fluoride and chloride do not interfere; silicate, borate, tungstate, and phosphate must be removed [2,3].

Method 2. Alloys containing between 20-35 mg of germanium are dissolved in aqua regia. After the germanium chloride is removed by distillation, the pH is adjusted to the neutral point (phenolphthalein), and then to a pH of 5.5 (p-nitrophenol). Mannitol is then added and the actual titration is begun using 0.02 N sodium hydroxide and phenol red as the indicator. The complex monoprotic germanium-mannitol acid is titrated to a definite red tint (pH 6.8). The end point may also be detected by means of a potentiometer or pH meter.

There is considerable saving in time in comparison to gravimetric procedures. The standard deviation in the analysis of a gold-germanium
alloy with 12.00% germanium is $\pm$ 0.19% [4].

Method 3. Germanium(IV) in samples containing 20-100 mg Ge/250
ml are determined by adjusting the pH of the sample to 5.0 and adding
an excess of pyrocatechol. The excess is then back-titrated potentiometrically to pH 5.0 with 0.1 N carbonate-free sodium hydroxide.
Arsenic(III) does not interfere. In the presence of antimony(III),
tin(IV), iron(III), or boron, the germanium is first separated by distillation from 6 N HCl before titration. The method has wide application in the analysis of germanium concentrates, intermediates in the
manufacture of germanium, etc. The error varies between $\pm$ 0.4% [5].

Complexometric Titration

Germanium solutions buffered with sodium acetate and sodium tetraborate are titrated amperometrically with standard pyrocatechol at
pH 7-9 at a dropping mercury electrode. The voltage applied is -1.7
V versus a saturated calomel electrode. The germanium must be freed
of interferences from heavy metals by extracting it as the tetrachloride with carbon tetrachloride. Arsenic(III), if present, is to be
oxidized to the pentavalent state with iodine [6].

Precipitation Titration

Germanium is titrated amperometrically with standard magnesium
sulfate solution at pH 10. The titration is carried out in a buffer
solution consisting of molar ammonium chloride and 0.5 molar ammonium
hydroxide plus 0.1% of gelatin. Arsenic(III), boron, and silicon do
not interfere. The sensitivity is 1×10^{-5} gm germanium per milliliter [7].

References

[1]. N. H. Furman, ed., *Standard Methods of Chemical Analysis*, 6th
     Ed., Vol. 1, D. Van Nostrand, Princeton, N. J., 1962, p. 467.

[2]. E. P. Bil'tyukova, *Stekle, Inform. Byul. Vses. Gos. Nauchn.-
     Issled. Inst. Stkla, 1963*(1), 11-15.

[3]. E. P. Bil'tyukova, *Stekle, Inform. Byul. Vses. Gos. Nauchn.-
     Issled. Inst. Stkla, 1963*(2), 72-75.

[4]. J. F. Reed, *Anal. Chem., 38*(8), 1085-1086 (1966).

[5]. E. Wunderlich and E. Goehring, *Z. Anal. Chem., 169*, 346-350
     (1959).

[6]. A. I. Zelyanskaya and N. V. Stashkova, *Zh. Analit. Khim., 16*,
     430-432 (1961).

[7]. R. M. Dranitskaya and E. V. Iashvili, *Zh. Analit. Khim.*, *19*,
      1031-1033 (1964).

TIN

Modern methods for the determination of tin include redox and
complexometric titrations.

Synoptic Survey

Classical Methods

Redox Titrations. Tin(II) is titrated with standard ferric
chloride in a hot solution [1].
Tin(IV) is reduced to tin(II) which is then titrated with standard iodine [2].

Contemporary Methods

Redox Titrations. Tin(II) is titrated with standard sodium vanadate in the presence of 3,3'-diphenylbenzidine and methylene blue as
the indicator [3].
Tin(II) is titrated in concentrated hydrochloric acid solution
with standard potassium ferricyanide using 3,3'-dimethylnaphthidine
or o-dianisidine as the indicator [4,5].
Tin(IV) together with cadmium and lead are determined in one
solution by controlled potential coulometry [6,7].
Tin(II) is determined indirectly by reaction with iodine produced
in the oxidation of KI with standard copper sulfate [8].

Complexometric Titration. Tin(II) in the presence of ammonium
fluoride and tartaric acid is titrated with standard EDTA using methylene blue as the indicator [9].

Outline of Recommended Contemporary Methods for Tin

Redox Titrations

Method 1. Tin(II) in samples ranging from 5-25 ml of 0.1-0.05
N solutions is titrated with standard sodium orthovanadate using 3,3'-
diphenylbenzidine and methylene blue as the indicator. The relative
error is less than 0.5% [3].

Method 2. Tin(II) is titrated in 6-9 M hydrochloric acid solution with standard potassium ferricyanide and 3,3'-dimethylnaphthidine
or o-dianisidine as the indicator.

$$SnCl_4^{-2} + 2\ Cl^- + 2\ Fe(CN)_6^{-3} \rightarrow SnCl_6^{-2} + 2\ Fe(CN)_6^{-4}$$

There is less than 0.2% difference between the values obtained by this

and the iodometric method. Lead, silver, zinc, and aluminum do not
interfere. However, substances which are oxidized by potassium ferricyanide must be absent. The entire analysis can be finished in only
a few minutes [4,5].

Method 3. In the absence of lead and cadmium, tin is readily
determined by controlled potential coulometric titration. The method
involves the reduction of tin(II) in an acidic bromide solution to the
amalgam and conclusion by coulometric anodic stripping. If a mercury
pool is used, it is possible to determine lead(II), cadmium(II), and
tin(IV) in the same solution. In this case, the lead and cadmium are
removed by reduction from an ammoniacal solution using a mercury cathode at -0.9 V. The amalgam is removed and preserved. Fresh mercury
is added to form a new working electrode, and the solution is acidified
and made 0.2 M with hydrobromic acid. Sufficient sodium bromide is
added to bring the concentration of NaBr to 3 M, and the tin is titrated as described above. The lead-cadmium amalgam is placed then in a
clean titration cell with sufficient 3 M sodium bromide and 0.25 M tartaric acid and both are reduced at -0.85 V and oxidized at -0.30 V.
After oxidation has been completed, the lead is reduced at -0.57 V and
oxidized at -0.30 V. Cadmium(II) does not reduce with the lead(II) at
-0.57 V and, thus, cadmium values can be found by difference [6,7].

Method 4. An indirect redox titration of tin is based on the
reaction of copper(II) with potassium iodide:

$$2 \ CuSO_4 \ + \ 4 \ KI \ \rightarrow \ Cu_2I_2 \ + \ 2 \ K_2SO_4 \ + \ I_2$$

The iodine is formed by adding a known volume of standard copper sulfate to excess potassium iodide. A definite volume of the unknown
stannous chloride solution is then added which brings about the reduction of an equivalent amount of the iodine. The remaining iodine
is titrated with standard thiosulfate. Carbon dioxide atmosphere is
not needed, however, it is recommended that the water used in making
up the solutions be free of dissolved oxygen. The percent error is
smaller than in many other methods. Percent errors are as follows:

Tin concn. (N)	Error (%)
.04289	0
0.02170	-1.1
0.01072	-2.3

Complexometric Titration

Tin(II) is titrated in the presence of ammonium fluoride and
tartaric acid at pH 5.5-6.0 with standard EDTA and methylthymol blue
as the indicator. Tin(IV) or antimony(III) do not interfere [9].

References

[1]. N. H. Furman, ed., *Standard Methods of Chemical Analysis*, 6th
      Ed., Vol. 1, D. Van Nostrand, Princeton, N. J., 1962, p. 1081.

[2]. ASTM Standards 1960, Chemical Analysis of Metals.

[3]. W. Wawrzyczek and K. Wisniewski, *Chem. Anal. (Warsaw)*, *10*(6),
      1287-1289 (1965).

[4]. H. Basinska and W. Rychcik, *Talanta*, *10*(12), 1299-1301 (1963).

[5]. H. Basinska and W. Rychcik, *Studia Soc. Sci. Torun Sect. B*, *5*(2),
      67-72 (1964).

[6]. W. M. Wise and J. P. Williams, *Anal. Chem.*, *37*(10), 1292-1294
      (1965).

[7]. W. M. Wise and D. E. Campbell, *Anal. Chem.*, *38*(8), 1979-1980
      (1966).

[8]. J. C. Gupta and S. P. Srivastava, *Z. Anal. Chem.*, *191*, 267-273
      (1962).

[9]. I. Dubský, *Collection Czechoslov. Chem. Communs.*, *24*, 4045-4048
      (1959).

LEAD

Methods for the determination of lead include redox, complexometric, and precipitation titrations.

Synoptic Survey

Classical Methods

Redox Titrations. Lead is determined by the addition of excess
standard potassium dichromate solution followed by back-titration with
standard ferrous sulfate solution. An alternate procedure involves
the addition of potassium iodide followed by titration with standard
sodium thiosulfate [1].
Lead is precipitated as the oxalate which in turn is separated,
decomposed with sulfuric acid and the liberated oxalic acid is titrated with standard potassium permanganate [2].

Precipitation Titration. Lead is precipitated as the ferrocyanide by titration with standard potassium ferrocyanide [3].
Lead is precipitated as the molybdate by titration with standard
ammonium molybdate [4].

Contemporary Methods

Complexometric Titrations. Lead alone or in the presence of phosphate and the alkaline earths is titrated with excess EDTA and the slight excess added is back-titrated with standard copper(II) using PAN as the indicator [5].

Lead is titrated amperometrically with standard (1,2-cyclohex-anedinitrilo)-tetraacetic acid, CDTA, in 2-propanol [6].

Lead in trace amounts is titrated photometrically with standard dithizone in chloroform [7].

Precipitation Titration. Lead is titrated in acetic acid with standard cadmium nitrate dissolved in acetic acid. Diphenylamine in the presence of potassium chromate is the indicator [8].

Outline of Recommended Contemporary Methods for Lead

Complexometric Titrations

Method 1. Samples containing 25 mg of lead are dissolved in nitric acid and are diluted to 50 ml, the pH is adjusted to pH 5 with a sodium acetate buffer. The alkaline earths as well as orthophosphate do not interfere. The lead sample is then titrated with 0.01 M EDTA until the original turbidity has disappeared and the solution is clear. The excess EDTA so required is back-titrated with 0.01 M copper(II) solution and PAN as the indicator after prior addition of 50 ml of ethanol. At the end point the color of the solution changes from a canary yellow to a persistent violet. Copper(II), nickel(II), mer-cury(II), iron(III), cadmium(II), zinc(II), and aluminum(III) inter-fere. The accuracy of the method is very good [5].

Method 2. Lead is titrated amperometrically with 0.1 M CDTA in 2-propanol which is 2 M with respect to ethanolamine. Since chelation is taking place in nonaqueous solutions, protons are released from CDTA. These are neutralized by the excess base favoring chelation and stabilizing the reaction. The average deviation is 1×10^{-5} (6 parts per 1000) for a lead concentration of 1.64×10^{-3} M [6].

Method 3. Lead in trace amounts is titrated spectrophotometri-cally at 510 mμ with standard dithizone dissolved in chloroform. Tin (II), thallium(I), copper(II), arsenic, cobalt(II), zinc, nickel(II), manganese(II), calcium, iron(II), iron(III), aluminum, phosphate, sul-fate, acetate, carbonate, and chloride do not interfere. Average recovery of lead in the presence of bismuth is 99.9% [7].

Precipitation Titration

Samples containing 0.02-0.10 gm lead in 1-10 ml acetic acid are titrated with 0.15 N cadmium nitrate dissolved in acetic acid. Di-phenylamine in the presence of potassium chromate acts as the indica-tor. The chromate ion oxidizes the diphenylamine. However, such oxidation occurs only after the initially formed lead chromate is converted to the insoluble lead nitrate. Titration is continued until the suspension changes from yellow to a dark green. Barium, mercury, bismuth, cadmium, zinc, and magnesium do not interfere [8].

References

[1].   N. H. Furman, ed., *Standard Methods of Chemical Analysis*, 6th Ed., Vol. 1, D. Van Nostrand, Princeton, N. J., 1962, pp. 566-567.

[2].   N. H. Furman, ed., *Standard Methods of Chemical Analysis*, 6th Ed., Vol. 1, D. Van Nostrand, Princeton, N. J., 1962, p. 564.

[3].   N. H. Furman, ed., *Standard Methods of Chemical Analysis*, 6th Ed., Vol. 1, D. Van Nostrand, Princeton, N. J., 1962, p. 563.

[4].   N. H. Furman, ed., *Standard Methods of Chemical Analysis*, 6th Ed., Vol. 1, D. Van Nostrand, Princeton, N. J., 1962, p. 565.

[5].   F. H. Blood and W. H. Nebergall, *Anal. Chem.*, *35*, 1089-1090 (1963).

[6].   P. Arthur and B. R. Hunt, *Anal. Chem.*, *39*(1), 95-97 (1967).

[7].   H. A. Jones and A. Szutka, *Anal. Chem.*, *38*(6), 779-787 (1966).

[8].   V. A. Bork and G. P. Fedukhina, *Tr. Mosk. Khim.-Tekhnol. Inst.* *51*, 250-253 (1966).

TITANIUM

Methods for the determination of titanium include redox and complexometric titrations.

Synoptic Survey

Classical Methods

Redox Titrations. Titanium(IV) is reduced to titanium(III) either with amalgamated-zinc Jones reductor or with aluminum foil, and titrated with standard ferric ammonium sulfate [1,2] or potassium permanganate [3].
Titanium(IV) is titrated with standard chromous ion solution [4].

Contemporary Methods

Redox Titrations. Titanium in the presence of iron is reduced in a Jones reductor. The solution is added to excess standard ferric alum solution. The iron(II) is titrated with standard potassium dichromate using barium diphenylamine sulfonate as the indicator.

Iron is determined separately in the sample by reduction in a silver reductor and titrating as before [5].

Titanium(IV) is reduced to the trivalent state with liquid zinc amalgam and titrated coulometrically with iron(III) [6].

The sample containing titanium(III) is added to excess standard ferric ion solution and the resulting ferrous ions are titrated with standard cerium(IV) solution and ferrous 1,10-phenanthroline as the indicator [7].

Complexometric Titrations. Titanium in the presence of niobium and tantalum is titrated with excess standard diaminocyclohexane tetra-acetic acid (DCyTA). The excess is back-titrated with standard copper solution in the presence of H_2O_2 and methyl calcein as the fluoro-chromic indicator [9].

Titanium(IV) is reduced with chromium(II) or vanadium(II) and then titrated photometrically with standard iron(III) solution [8].

Outline of Recommended Contemporary Methods for Titanium

Redox Titrations

Method 1. The solution containing titanium(IV) and iron(III) is passed through a Jones reductor into a solution of ferric alum in excess. Titanium(III) reduces an equivalent amount of ferric to ferr-ous ions which, in turn and together with the ferrous ions from iron (III) in the sample, are titrated with standard dichromate using bar-ium diphenylamine sulfonate as the indicator. The iron(III) value of the original sample is found independently by reduction in a silver reductor followed by titration with standard dichromate. The method is simple, fast, and accurate. Samples containing small amounts of titanium can also be analyzed successfully. Chromate does not inter-fere [5].

Method 2. Titanium(IV) is reduced by means of the Somey reductor which is charged with liquid zinc amalgam. The titanium(III) so formed is titrated by coulometrically generated iron(III). The equivalence point is found biamperometrically or bipotentiometrically. Samples containing 1-1000 γ of titanium in 10 ml can be determined. In the concentration range of 1×10^{-3} to $1 \times 10^{-4}\%$ titanium, the average error is $\pm$ 3.8% [6].

Method 3. Titanium(III) from magnesia-titania refractories is dissolved in boiling sulfuric acid containing ferric sulfate under an atmosphere of carbon dioxide. The equivalent amount of ferrous ions formed are titrated with standard ceric sulfate solution using ferrous 1,10-phenanthroline as the indicator. At the end point the color change is from orange to colorless. The method is rapid. Elements which are oxidized by cerium(IV) must be absent [7].

Method 4. Titanium(IV) is reduced by a small excess of chrom-ium(II) or vanadium(II) in the presence of acetylacetone to form the colored titanium(III) complex. The titanium(III) is then titrated

under nitrogen with an iron(III) solution. As equilibrium is established rather slowly at room temperature, all titrations are best performed at 50°. Titrations are carried out spectrophotometrically at
490 mµ. Metals which are often present either do not complex with the
acetylacetone, or their complexes have a low molar absorption at the
maximum absorbance of the titanium(III) complex. Standard deviation
in the absence of other metals is 0.9% for the determination of 356
µg of titanium. A systematic error of -2.8% is noted if the iron(III)
solution is not deoxygenated; this is reduced to -0.3% by deoxygena-
tion. Iron, chromium, nickel, vanadium, cobalt, manganese, and aluminum do not interfere. Large amounts of nickel, chromium, and especially cobalt decrease the accuracy. Copper should be removed by prereduction with an amalgamated reductor. Molybdenum, tungsten, and
niobium interfere [8].

Complexometric Titration

Samples containing 1-10 mg in the presence of niobium and tantalum are determined by adding an excess of standard DCyTA. The pH is
adjusted to a pH 5-5.5 by the addition of ammonia, 1 ml of 30% hydrogen
peroxide is added, and the excess DCyTA is back-titrated with 0.05 M
copper(II) sulfate. Methyl calcein or methyl calcein blue is used as
the metalfluorochromic indicator; a standard UV titration assembly is
used. Fluorescence is quenched at the end point. With samples having
a tantalum to niobium ratio greater than 5:1, methyl calcein blue
gives better results; however, in the presence of high concentrations
of niobium, methyl calcein is the indicator of choice. Cemented car-
bides can easily be analyzed by this method after removal of the
tungsten and cobalt. The average error over 20 titration is 0.3% [9].

References

[1]. J. A. Rahm, *Anal. Chem.*, *24*, 1832 (1952).

[2]. N. H. Furman, ed., *Standard Methods of Chemical Analysis*, 6th
Ed., Vol. 1, D. Van Nostrand, Princeton, N. J., 1962, pp. 1104-
1105.

[3]. B. A. Shippy, *Anal. Chem.*, *21*, 698 (1949).

[4]. J. J. Lingane, *Anal. Chem.*, *20*, 797 (1948).

[5]. J. P. R. Fonseka and N. R. de Silva, *Anal. Chem.*, *36*(2), 437-
438 (1964).

[6]. Z. Slovak and M. Přibyl, *Z. Anal. Chem.*, *211*(4), 247-253 (1965).

[7]. Y. Su, *Anal. Chem.*, *38*(1), 129-130 (1966).

[8]. W. E. Van der Linden and G. den Beef, *Anal. Chim. Acta*, *38*(4),
517-522 (1967).

[9]. E. Lassner and R. Scharf, *Chemist-Analyst, 51,* 49 (1962).

ZIRCONIUM AND HAFNIUM

Because of the chemical similarities of zirconium and hafnium, there is, at the present time, no titrimetric method available for the determination of one of these two elements in the presence of the other. Hence, in a determination of "zirconium," one determines a mixture of zirconium and hafnium, of which hafnium usually accounts for less than 2% of the total. Zirconium/hafnium ratios can be determined by density measurements of the mixed oxides or by indirect analysis [1].

The best modern methods for the determination of zirconium or hafnium involve complexometric titrations.

Synoptic Survey

Classical Methods

Acid-Base Titrations. Zirconium salts are hydrolyzed and the liberated acid is titrated with standard base [2].

Zirconium hydroxide is titrated with standard hydrochloric acid after the addition of excess sodium fluoride to form the sodium fluozirconate [3].

Complexometric Titration. Zirconium is titrated at pH 1.3-1.5 with standard EDTA using Eriochrome cyanine RC, chromazurol S, or alizarol cyanone RC as the indicator [4].

Precipitation Titration. Zirconium is determined by adding excess standard potassium phosphate followed by back-titration with standard bismuth perchlorate using thiourea as an external indicator [5].

Contemporary Methods

Complexometric Titration. Zirconium and/or hafnium may be titrated in a nitric acid solution with EDTA using xylenol orange as the indicator [6].

Zirconium and/or hafnium may be titrated at pH 0.1-1.0 with standard EDTA using Stilbazogall II as the indicator [7].

Zirconium and/or hafnium may be determined in 5 M hydrochloric acid by titration with standard EDTA using catechol violet as the indicator [8].

Outline of Recommended Contemporary Methods for Zirconium/Hafnium

Complexometric Titrations

Method 1. Samples containing not more than 9 mg zirconium and

contained in 200 ml of solution which is 3 N in nitric acid are boiled
for five minutes to effect maximum depolymerization of the zirconium
polycations. The acid concentration is then lowered to 1 N, and the
zirconium is titrated in hot solution (above 90°) with 0.05 M EDTA
using xylenol orange as the indicator. The color change at the end
point is from pink to lemon yellow. Values agree favorably with those
obtained by gravimetric analysis. Fluoride and phosphate even in
traces interfere. When the above sample size of zirconium is used,
iron and titanium up to 181.5 mg and 155.7 mg, respectively, have no
effect; bismuth when present in excess of 62.70 mg interfered. For
samples containing 91.16 mg zirconium, the amount zirconium found is
91.47 mg or a difference of + 0.31 mg [6].

Method 2. Zirconium and/or hafnium may be titrated at pH 0.1-
1.0 with standard EDTA using Stilbazogall II, stilbene-2,2'-disulfonic
acid-4,4'-bis[(azo-1")-2"carboxy-4",5",6"-tri-hydroxybenzene] as a
selective indicator. The zirconium-Stilbazogall II complex shows a
maximum absorbance at 500 mμ. At the end point the color change is
from violet to yellow. Magnesium and aluminum ions, even in large
amounts, do not interfere. In small amounts, cadmium, zinc, titanium
(IV), thorium(IV), uranyl, cobalt, and nickel do not interfere. Interferences are noted from fluoride, phosphate, molybdate, tungstate, and
iron(III) [7].

Method 3. Zirconium and/or hafnium may be titrated in 5 M hydrochloric acid with 0.01 M EDTA using catechol violet as the indicator.
The complex formed contains a 1:1 molar ratio of EDTA to zirconium.
The color change at the end point is from blue to yellow. The method
is fast and shows high precision and accuracy. The use of catechol
violet is to be preferred to other indicators currently being used [8].

References

[1]. R. B. Hahn, in *Treatise on Analytical Chemistry*, I. M. Kolthoff
     and P. J. Elving, eds., Interscience, N. Y., Part II, Vol. 5,
     1961, pp. 61-138.

[2]. D. Bezier, *Chim. Anal.*, *36*, 175 (1954).

[3]. T. Sawaya and M. Yamashita, *J. Chem. Soc. Japan, Pure Chem.
     Sect.*, *72*, 256 (1951).

[4]. A. Musil and M. Theis, *Z. Anal. Chem.*, *144*, 427 (1955).

[5]. P. R. Subbaraman and K. S. Rejan, *J. Sci. Ind. Research (India)*,
     13B, 31 (1954).

[6]. B. C. Sinka and S. D. Gupta, *Analyst*, *92*, 558-564 (1967).

[7]. A. I. Cherkesov, Y. V. Pushinov, and V. S. Tonkoshkurov, *Zh.
     Analit. Khim.*, *20*(4), 459-461 (1965).

[8].   V. Baran and M. Tympl, *Collection Czech. Chem. Commun.*, *29*(9), 2250-2251 (1964).

GROUP V

Nitrogen	Bismuth
Phosphorus	Vanadium
Arsenic	Niobium
Antimony	Tantalum

NITROGEN

In this section only the titrimetry of nitrogen present as azide ion, hydroxylamine, hydrazine, ammonia, nitrite ion, nitrate ion, and nitrosyl chloride will be discussed [1].

Synoptic Survey of Azide Ion Determinations

Methods for the determination of azide ion include acid-base and redox titration.

Classical Methods

Redox Titration. Azide ion is determined by oxidation to nitrogen. Titrants are: cerate [2], permanganate [3], iodine [4], and nitrite [5].

Contemporary Methods

Acid-Base Titration. Azide ion is determined by measurement of the hydrogen ion consumed during oxidation by nitrite [6].

Redox Titration. The classic method of oxidation of azide with nitrite is improved by the use of an internal, reversible indicator, Ferrocyphen (dicyano-bis(1,10-phenanthroline)iron(II) [7].

Outline of Recommended Contemporary Methods for Azide Ion

Acid-Base Titrations

Method 1. Azide ion is determined by measurement of the hydrogen ion consumed during oxidation by nitrite according to the equation:

$$2 \text{ H}^+ + \text{N}_3^- + \text{NO}_2^- \rightarrow \text{H}_2\text{O} + \text{N}_2\text{O} + \text{N}_2$$

The pH of samples containing 0.5-1.5 mmole of azide are adjusted
to phenolphthalein end point. A 20% excess of 0.1 M perchloric acid
is added followed by an excess of about 1.0 M sodium nitrite. After
5-10 seconds stirring, the mixture is titrated with 0.1 M sodium hydroxide to a phenolphthalein end point. Two hydrogen ions are consumed
by each azide oxidized. Large amounts of chloride, thiocyanate, nitrate, and perchlorate do not interfere. Silver azide may be determined
after metathesis with sodium thiocyanate; lead azide is determined by
treatment with perchloric acid and distillation of the hydrazoic acid.
Palladium azide requires the addition of chloride as a complexing
agent and treatment with perchloric acid in order to recover by distillation all of the hydrazoic acid. Metal ions which hydrolyze at
pH 8-9 must be absent. The titration of 0.01 M azide is about the
limit of application. The relative standard deviation is 0.1% for
1.502 mmole and 0.5% for 0.500 mmole of sodium azide [6].

Method 2. Azide ion is titrated with 0.1 M sodium nitrite in
hydrochloric acid solution at 0° using ferrocyphen as the internal
refersible indicator.

$$[\text{Fe(phen)}_2(\text{CN})_2] \cdot \text{H}^+ + \text{HONO} \rightarrow [\text{Fe(phen)}_2(\text{CN})_2]^+ + \text{H}_2\text{O} + \text{NO}$$

    Ferrocyphen   Ferricyphen
   (orange-yellow)   (violet)

This is an important improvement over the classic approach which
suffered from the lack of a suitable indicator. Potentiometric end-
point determination is not feasible because of slow potential drifts;
dissolved oxygen gives low results. The titration can be finished in
8-10 minutes. In the assay of relatively pure sodium azide, a precision of 0.2% is obtained [7].

References for Azide Ion

[1]. A. J. Clear and M. Roth, in *Treatise on Analytical Chemistry*,
      I. M. Kolthoff and P. J. Elving, eds., Interscience, Part II,
      Vol. 5, 1961, pp. 291-292.

[2]. T. W. Arnold, *Anal. Chem.*, *17*, 215 (1945).

[3]. J. H. Van der Meulen, *Rec. Trav. Chim.*, *67*, 600 (1948).

[4]. F. Feigl and E. Chargav, *Z. Anal. Chem.*, *74*, 376 (1928).

[5]. I. M. Kolthoff and R. Belcher, *Volumetric Analysis*, Vol. III,
      Interscience, N. Y., 1957, pp. 660-662.

[6]. R. G. Clem and E. H. Hufmann, *Anal. Chem.*, *37*, 366-368 (1965).

[7]. A. A. Schilt and J. W. Sutherland, *Anal. Chem.*, *36*(9), 1805-
      1807 (1964).

Synoptic Survey of Hydrazine Determinations

Titrimetric methods for the determination of hydrazine involve redox procedures [1].

Classical Methods

Redox Titration. Hydrazine is titrated with iodine at pH 7.0-7.4 [1,2], with excess standard iodate in concentrated hydrochloric acid solution followed by the addition of potassium iodide; the liberated iodine is titrated with thiosulfate [2], with standard bromate [3], and with permanganate [4].

Contemporary Methods

Redox Titrations. Hydrazine dihydrochloride is titrated potentiometrically in hydrochloric acid solution with standard cerium(IV) [5].

Hydrazine and hydroxylamine may be determined in a mixture of the two by titration with ferricyanide [6].

Outline of Recommended Contemporary Methods for Hydrazine

Redox Titrations

Method 1. Hydrazine dihydrochloride is titrated potentiometrically with ammonium hexanitratocerate in the presence of potassium bromide as a catalyst. The potential is measured using platinum electrodes versus SCE. The end point may also be obtained visually using either α-naphthoflavone or p-ethoxyl-chrysoidine as the indicator; sulfate ion interferes. Relative standard deviation of the mean is about 0.2% [5].

Method 2. Although both hydrazine and hydroxylamine are strong reducing agents, they may be determined in the presence of each other by titration with standard potassium ferricyanide in the presence of zinc sulfate. Two titrations are required. In the first titration, an aliquot of the mixture containing hydrazine and hydroxylamine is treated with a known excess of standard ferricyanide solution followed by the addition of a borax-boric acid buffer (pH 8.9). After 15-20 minutes, the excess of ferricyanide is titrated iodometrically, or the ferrocyanide formed is titrated with standard ceric sulfate. The reactions occurring are:

$$N_2H_4 + 4K_3Fe(CN)_6 + 4KOH \rightarrow N_2 + 4K_4Fe(CN)_6 + 4H_2O$$

$$2NH_2OH + 2K_3Fe(CN)_6 + 2KOH \rightarrow N_2 + 2K_4Fe(CN)_6 + 4H_2O$$

In the second titration, a known volume of the ferricyanide solution is titrated hot in the presence of zinc sulfate and at a total alkalinity of 0.5-0.8 N with the mixed solution of hydrazine and hydroxylamine. The end point is shown by the appearance of a white precipitate. The reactions occurring are:

$$N_2H_4 + 4 K_3Fe(CN)_6 + 4 KOH + 6 ZnSO_4 \rightarrow N_2 + 2 Zn_3K_2[Fe(CN)_6]_2$$
$$+ 6 K_2SO_4 + 4 H_2O$$

$$NH_2OH + 6 K_3Fe(CN)_6 + 6 KOH + 9 ZnSO_4 \rightarrow HNO_3 + 3 Zn_3K_2-$$
$$[Fe(CN)_6]_2 + 9 K_2SO_4 + 4 H_2O$$

From the volumes of ferricyanide used by the same quantity of the mixture in both titrations, the weights of hydrazine and hydroxylamine are calculated. The error is $\leq$ 2% [6].

References for Hydrazine

[1]. L. F. Audrieth and B. A. Ogg, *The Chemistry of Hydrazine*, Wiley, N. Y., 1951, p. 160.

[2]. W. C. Bray and E. J. Cuy, *J. Amer. Chem. Soc.*, *46*, 858 (1924).

[3]. L. F. Audrieth and B. A. Ogg, *J. Amer. Chem. Soc.*, *46*, 162 (1924).

[4]. I. M. Issa and R. M. Issa, *Chemist-Analyst*, *45*, 40 (1956).

[5]. G. G. Rao and P. V. K. Rao, *Talanta*, *11*(11), 1489-1496 (1964).

[6]. B. R. Sant, *Anal. Chim. Acta*, *20*, 271-275 (1959).

Synoptic Survey of Hydroxylamine Determinations

Methods for the determination of hydroxylamine involve redox titration.

Classical Methods

Redox Titration. Hydroxylamine may be determined by oxidation by a variety of reagents such as: potassium bromate [1,2], titanium(III) [3], ferric ammonium sulfate [4], or ceric sulfate [5].

Contemporary Methods

Redox Titration. Hydroxylamine is determined by titration with ferric ion, and the iron(II) formed is determined with electrogenerated cerium(IV) [6].

Outline of Recommended Contemporary Methods for Hydroxylamine

Redox Titrations

Method 1. see Hydrazine, p. 95, for the determination of hydroxylamine in the presence of hydrazine.

Method 2. Hydroxylamine is determined by titration with standard ferric ammonium sulfate in hot 1 N sulfuric acid. The iron(II) formed is determined with cerium(IV) electrogenerated at constant current of 9.80 mA [6].

References for Hydroxylamine Determinations

[1]. A. Kurtenacker and J. Wagner, *Z. Anorg. Chem.*, *120*, 261 (1921).

[2]. N. H. Furman and J. H. Flagg, *Ind. Eng. Chem. Anal. Ed.*, *12*, 738 (1940).

[3]. W. C. Brag, M. E. Simpson, and A. A. MacKenzie, *J. Amer. Chem. Soc.*, *41*, 1362 (1919).

[4]. N. H. Furman, ed., *Standard Methods of Chemical Analysis*, 6th Ed., Vol. 1, D. Van Nostrand, Princeton, N. J., 1962, p. 759.

[5]. S. R. Cooper and J. B. Morris, *Anal. Chem.*, *24*, 1360 (1962).

[6]. T. Takahashi and H. Sakurai, *Rep. Inst. Ind. Sci. Univ. Tokyo*, *13*(1), 1-37 (1963).

Synoptic Survey of Ammonia and Ammonium Salt Determinations

Methods for the determination of ammonia and ammonium salts include acid-base, redox, and complexometric titrations.

Classical Methods

Acid-Base Titrations. Ammonia is titrated with a standard strong acid using methyl red as the indicator [1].
Ammonium salts are treated with excess strong base, and the liberated ammonia is distilled with steam and collected in standard strong acid, or in a saturated solution of boric acid. The excess acid is then back-titrated in the usual fashion [2].

Contemporary Methods

Redox Titration. Ammonium ion is oxidized with excess standard hypobromite and the excess is back-titrated with standard hydrogen peroxide [3].

Complexometric Titrations. Ammonium ion is allowed to react

with the mercury(II)-EDTA complex and the freed hydrogen ion is titrated with standard base [4].

   Ammonium ion is precipitated with tetraphenylboron; excess mer-
cury(II) EDTA complex is added, and the liberated EDTA is titrated
with standard magnesium solution [4,5].

   Ammonium ion is precipitated with tetraphenylboron which is dissolved in acetone and titrated with standard silver nitrate solution
using variamine blue acetate as the indicator [6].

   Outline of Recommended Contemporary Methods for Ammonium Ion

Redox Titration

   Ammonium ion is determined by the addition of excess standard
hypobromite ion. The excess is back-titrated with standard hydrogen
peroxide. The hypobromite oxidizes the ammonium ion forming nitrogen
according to the equation:

$$2\ NH_4^+ + 3\ OBr^- + 2\ OH^- \rightarrow N_2 + 3\ Br^- + 5\ H_2O$$

The method is well suited for control purposes and provides an accuracy of about 1% [3].

Complexometric Titrations

   Method 1. Ammonium ion is determined in a mixture with alkali
metal ions and free ammonia by titration of the acid liberated by the
reaction between mercury(II)-EDTA and the ammonium ion. Ammonia and
ammonium ions react with the mercury(II)-EDTA complex according to the
equations:

$$HgY^{-2} + NH_3 \rightleftarrows HgYNH_3^{-2}$$

$$HgY^{-2} + NH_4^+ \rightleftarrows HgYNH_3^{-2} + H^+$$

   The method is practically free of all anion interference except
phosphate. Accurate results are obtained in both the semimicro and
the micro ranges. The log of the stability constant for $HgYNH_3^{-2}$ is
equal to 6.4. The titration is carried out using as the indicator
bromothymol blue and a red screening dye such as ponceau PXO(C179) or
acid red 220%(C131) (Williams Hounslaw Ltd., Middlesex, England). The
error is less than 0.7% [4].

   Method 2. Ammonium ion (and/or potassium ion) may be precipitated with tetraphenylboron. The precipitate is removed, and dissolved
in N,N-dimethylformamide. Excess mercury(II)-EDTA is added and the
liberated EDTA is titrated with standard magnesium solution according
to the equation:

$$4\ HgY^{-2} + (Ph)_4B^- + (4n)H_2O \rightarrow H_3BO_3 + 4\ PhHg^+$$
$$+ 4\ H_nY^{n-4} + (4n-3)OH^-$$

Eriochrome Black T is used as the indicator. Each mole of ammonia or potassium ions results in the liberation of 4 moles of EDTA. The error varies between 0.51% for 7.216 mg and 0.83% for 0.520 mg ammonium ion respectively [4,5].

Method 3. Ammonium ion is precipitated with the tetraphenylboron ion. The precipitate is dissolved in acetone and titrated with standard silver nitrate solution using variamine blue acetate as the indicator. Trivalent nitrogen in organic compounds can be determined after a Kjeldahl digestion. The standard deviation is ± 0.19% for samples of ammonium sulfate weighing 130 mg [6].

References for Ammonia and Ammonium Salts

[1]. N. H. Furman, ed., *Standard Methods of Chemical Analysis*, 6th
      Ed., Vol. 1, D. Van Nostrand, Princeton, N. J., 1962, p. 744.

[2]. N. H. Furman, ed., *Standard Methods of Chemical Analysis*, 6th
      Ed., Vol. 1, D. Van Nostrand, Princeton, N. J., 1962, p. 745.

[3]. L. Erdey and J. Inczédy, *Z. Anal. Chem.*, *166*, 410-417 (1959).

[4]. F. S. Sadek and C. N. Reilly, *Anal. Chem.*, *31*, 494-498 (1959).

[5]. H. Flaschka and F. Sadek, *Chemist-Analyst*, *47*, 30 (1958).

[6]. L. Erdey, K. Vigh, and L. Pólos, *Talanta*, *3*, 1-5 (1959).

Synoptic Survey of Nitrite Determinations

Methods for the determination of nitrite ion include acid-base and redox titration.

Classical Methods

Acid-Base Titration. Nitrite and nitrogen trioxide are titrated with standard permanganate to nitrate and nitrogen pentoxide respectively [1].

Contemporary Methods

Acid-Base Titration. Alkali nitrites are determined by high frequency titration with 0.1 N hydrochloric acid in 30% ethyl alcohol solution [2].

Redox Titrations. Nitrite is titrated amperometrically with standard sulfamic acid, or standard ceric ion [3], or permanganate [4].
Nitrite is titrated potentiometrically with standard lead tetraacetate [5].
Small amounts of nitrite may be titrated potentiometrically with standard solutions of a primary aromatic amine [6].

Outline of Recommended Contemporary Methods for Nitrite Ion

Acid-Base Titration

Soluble nitrites are determined by high frequency titration in 30% ethyl alcohol with 0.1 N hydrochloric acid. The end point is measured at 130 Mc/second in 130 ml of solution. Samples containing from 10-50 mg of sodium nitrite may be determined in the presence of 50-200 mg of potassium nitrate and 50-100 mg of ammonium chloride. No inflection of the curve is noted in aqueous solution. The relative error is $\pm$ 2% [2].

Redox Titrations

Method 1. Nitrite ion may be determined in 6 x 10^{-4} M concentration by amperometric titration in 0.05 M sulfuric acid. A rotating platinum microanode is used with either standard sulfamic acid or ceric solution as the titrant. The titration is carried out at a potential of + 1.05 V versus SCE. While the titration with ceric ion is less precise than using standard sulfamic acid, lower nitrite concentrations can be determined. The titration with sulfamic acid is free from bias and the standard deviation is about 1% [3].

Method 2. Low concentrations of nitrite are titrated amperometrically in 0.1 M sulfuric acid with 0.2 N potassium permanganate at a potential of + 1.0 V. An "L" type titration curve is found. A rotation platinum wire electrode is used which is short-circuited in 0.1 M perchloric acid against a SCE and depolarized at + 1.0 V in a solution of concentration similar to that titrated. The precision is good but the procedure is by no means specific for nitrite [4].

Method 3. Nitrite ion is determined by quantitative oxidation to nitrate with lead tetraacetate. The end point is determined potentiometrically using a platinum indicating electrode. Samples which contain 0.1-1.0 mg nitrite/20 ml and which have been made 1 M or 0.1 M respectively in sodium chloride are titrated with 0.1 M lead tetraacetate. The end point is indicated by a change in potential of 200 mV/0.01 ml of titrant at about 795 mV versus SCE. The error for 2-3 mg of nitrite ion is $\pm$ 0.17% and for samples containing 0.1 mg of nitrite, the error is $\pm$ 0.58% [5].

Method 4. Samples containing from 0.7-1.5 mg of sodium or potassium nitrite are determined by diazotization titrimetry with a standard solution of primary amine. All details are missing in the original article. Sodium nitrate does not interfere. The relative accuracy is reported to be $\pm$ 0.5-1.0% [6].

References for Nitrite Ion

[1]. J. S. Laird and T. C. Simpson, *J. Amer. Chem. Soc.*, *41*, 524 (1919).

[2]. I. Krausz and H. A. Endroi, *Magyar Kêm. Folyôirat*, *67*, 454-455 (1961).

[3]. J. T. Stock and R. G. Bjork, *Talanta*, *11*(2), 315-319 (1964).

[4]. J. T. Stock and R. G. Bjork, *Microchem. J.*, *6*, 219-224 (1962).

[5]. A. Morales and J. Zyka, *Collection Czech. Chem. Commun.*, *27*, 1029-1030 (1962).

[6]. S. Kh. Dzottsoti, *Azerbaidzhan Khim. Zhur.*, *1961*, *1*, 85-96.

Synoptic Survey of Nitrate Determinations

Methods for the determination of nitrate include acid-base, redox, and complexometric titrations.

Classical Methods

Acid-Base Titration. Nitrate is determined by reduction in a basic solution with Devarda's alloy. The ammonia produced is distilled into excess standard strong acid or into excess solution of boric acid and titrated with standard base [1].

Redox Titrations. Nitrate is determined by reduction with manganese(II) chloride followed by distillation of the reaction products, nitrogen dioxide, chlorine, etc. into potassium iodide solution. The liberated iodine is then titrated with standard thiosulfate [2].

Nitrate is reduced with excess standard ferrous sulfate and the excess is back-titrated with standard potassium dichromate [3].

Contemporary Methods

Acid-Base Titration. Nitrates of Group III and higher are titrated potentiometrically with KOH. Nitrates of Groups I and II are passed through an ion exchange resin, and the nitric acid in the effluent is titrated with tetraethylammonium hydroxide [4].

Complexometric Titration. Nitrate is reduced to ammonia. The ammonia is precipitated as ammonium tetraphenylboron which is dissolved in acetone and titrated with standard silver ion using variamine blue acetate as the indicator [5].

Precipitation Titration. Nitrate is determined by amperometric titration with lead acetate in acetic acid solution [6].

Outline of Recommended Contemporary Methods for Nitrate Ion

Acid-Base Titration

Samples containing 5 mg of the metal nitrate are dissolved in methanol, and diluted to 20 ml with methanol or acetone. If the sample contains nitrates of Group III or higher Groups of the periodic table, the nitrate is determined by potentiometric titration with standard potassium hydroxide. Nitrates of Group I and Group II are passed through a cation exchange resin, and the nitric acid in the effluent is titrated with tetraethylammonium hydroxide. Nitrate-nitrite mixtures are analyzed by combining this procedure for nitrate with a titration of nitrite using standard solutions of perchloric acid. The error is $\pm$ 1.5% in the analysis of nitrate-nitrite mixtures [4].

Complexometric Titration

Nitrate ion is determined by reduction with metallic iron in the presence of nickel sulfate or with Devarda's alloy in basic solution. The ammonia produced by such reduction is then distilled and titrated in the normal fashion, or precipitated as ammonium tetraphenylboron. In the latter case, the ammonium tetraphenylboron is dissolved in acetone and titrated with standard silver nitrate using variamine blue acetate as the indicator. The method has rather wide application and may be applied to the determination of nitrates even when accompanied by organic material. It may also be used to determine the total nitrogen content of fertilizers and in the analysis of nitro-, azo- and diazocompounds. The accuracy is excellent; standard deviation is $\pm$ 0.25% [5].

Precipitation Titration

Nitrate is determined by amperometric titration with lead acetate solutions in acetic acid which is 0.1 M in lithium acetate. The titration is carried out at -1.0 V using a dropping mercury electrode versus a mercury pool. The end point may also be determined visually by adding potassium chromate and diphenylamine to the standard lead acetate titrant [6].

References for Nitrate

[1]. N. H. Furman, ed., *Standard Methods of Chemical Analysis,* 6th Ed., Vol. 1, D. Van Nostrand, Princeton, N. J., 1962, pp. 748-752.

[2]. N. H. Furman, ed., *Standard Methods of Chemical Analysis,* 6th Ed., Vol. 1, D. Van Nostrand, Princeton, N. J., 1962, pp. 753-754.

[3]. I. M. Kolthoff and B. Moskovitz, *J. Amer. Chem. Soc., 55,* 1454 (1933).

[4]. A. M. Birun, K. A. Komarova, E. K. Kreshkova, and A. N. Yarovenko, *Izv. Vyssh. Ucheb. Zaved. Khim. Khim. Tekhnol., 9*(4), 546-559 (1966).

[5]. L. Erdey, L. Polos, and A. Gregorowicz, *Talanta, 3,* 6-13 (1959).

[6]. A. P. Kreshkov, V. A. Bork, L. A. Shvyrkova, and M. I. Aparsheva, *Zavodsk. Lab., 32*(1), 10-12 (1966).

Synoptic Survey of Nitrosyl Chloride Determinations

Nitrosyl chloride is determined by redox titration.

Classical Methods

None of importance.

Contemporary Methods

Redox Titration. Nitrosyl chloride is determined iodometrically at pH 3-3.5 [1].

Outline of a Recommended Contemporary Method for Nitrosyl Chloride

Redox Titration

Nitrosyl chloride is determined iodometrically based on the following reaction:

$$2 \text{ NOCl} + 2 \text{ KI} \rightarrow I_2 + 2 \text{ KCl} + 2 \text{ NO}$$

The titration is quantitative only at pH 3-3.5, and is carried out under a nitrogen atmosphere. Other oxidants which oxidize potassium iodide must be absent. The liberated iodine is titrated with standard solution of sodium thiosulfate [1].

References for Nitrosyl Chloride

[1]. V. K. Bukina, M. F. Prokopeva, and S. E. Yartudakis, *Doklady Akad. Nauk. Uzbek, S.S.R., 1961, 9,* 20-22.

PHOSPHORUS

Methods involved in the determination of phosphorus include acid-base, redox, complexometric, and precipitation titrations of phosphorus anions.

Synoptic Survey

Classical Methods

Acid-Base Titration of Acids of Phosphorus. Every acid of phosphorus contains one strongly acidic hydrogen for each phosphorus atom.

Determination may be carried out by titration with standard base to the first end point near pH 4.5 [1].

Redox Titration of Hypophosphite and Phosphite. Samples are titrated involving oxidation to phosphate with standard bromate [2], iodine [3], hypobromite [4], or vanadate solution [5].

Acid-Base Titration of Orthophosphate. Orthophosphate is precipitated as ammonium phosphomolybdate, $(NH_4)_3PO_4 \cdot 12MoO_3$. The precipitate is dissolved in excess standard sodium hydroxide and back-titrated with standard nitric acid using phenolphthalein as the indicator [6].

Contemporary Methods

Redox Titrations of Hypophosphite. Hypophosphite is determined by potentiometric titration using a palladium(II) solution [7].
Hypophosphite is determined by oxidation with standard ceric solution in the presence of a mixed catalyst containing $Mn(NO_3)_2$ and $AgNO_3$ [8].

Redox Titrations of Phosphite. Phosphite is determined by titration with excess standard iodine solution containing small amounts of potassium iodide. The excess is back-titrated with standard arsenite or hydrazine sulfate [9].
Phosphite is oxidized with excess standard ceric solution in the presence of a mixed catalyst containing $Mn(NO_3)_2$ and $AgNO_3$. The excess is back-titrated with ferrous ion [10].
Phosphorus and hypophosphorus acids are determined in the presence of each other by oxidizing the latter with silver perchlorate. The sum of both acids is determined bromatimetrically [11].

Precipitation Titrations of Orthophosphate. Orthophosphate is determined by potentiometric titration with silver ion at a silver electrode [12] or by potentiometric titration in ethanol which is 0.1 M in sodium acetate with standard silver nitrate or with silver obtained by generation from a silver anode [13].
Orthophosphate may be titrated with standard bismuth(III) solution using xylenol orange as the indicator [14], or by titration with standard cerous ion solution in the presence of Eriochrome Black T as the indicator [15].
Orthophosphate may be precipitated as quinoline phosphomolybdate which in turn is titrated with standard sodium hydroxide [16,17].
Orthophosphate is precipitated as the phosphomolybdate which is removed by extraction, decomposed with alkali, and titrated, following reduction with silver reductor, with cerium(IV) sulfate [18].

Outline of Recommended Contemporary Methods for Phosphorus Anions

Redox Titrations of Hypophosphite

Method 1. Samples containing 5 gm/liter of sodium hypophosphite are titrated potentiometrically at pH 2.8 into a vessel containing a

known amount of palladium(II) salt solution. For the titration, a
palladium indicator electrode versus SCE is used. Palladium ion is
reduced during the titration to the metal, and hydrogen is liberated
on the surface of the palladium electrode. The potential of this
electrode increases about 350 mV at the end point. The titration is
carried out at a temperature of 80-90°. Phosphite ion, hydrazine, and
ions oxidized by palladium such as sulfite interfere. The precision
is 1.0% [7].

Method 2. Samples containing 5-35 mg phosphite ion are oxidized
in 1 F sulfuric acid solution by the addition of 70-150% excess of
standard cerium(IV) ion. A small amount of a mixed catalyst consisting
of manganous nitrate and silver nitrate is added to promote the reaction. After heating at 100° for 7-10 minutes or more, the excess ceric
ion is back-titrated with standard iron(II) solution using ferroin as
the indicator. The standard deviation is ± 0.32% [8].

Redox Titrations of Phosphite

Method 1. By the use of a lower potassium iodide content of 1.7%
in a 0.1 N iodine solution, the usual hinderance of phosphite oxidation by iodide is avoided, and the iodimetric determination of phosphorous acid is considerably increased. Samples containing not more
than 1 meq are titrated in a buffered solution by the addition of at
least a twofold excess of 0.1 N iodine solution containing 1.7% potassium iodide. The excess iodine is back-titrated with standard
arsenite or hydrazine sulfate after acidification with acetic acid.
Increasing the concentration of the buffer, sodium acid carbonate,
disodium hydrogen phosphate, or sodium tetraborate also increases the
rate of phosphite oxidation [9].

Method 2. Samples containing 10-100 mg phosphate in 1 F sulfuric
acid are oxidized by the addition of 70-150% excess standard ceric sulfate. A small amount of a mixed catalyst composed of manganous nitrate
and silver nitrate is added to increase the rate of the reaction. After
heating for 7-10 minutes or more at 100°, the reaction is quenched by
cooling and the addition of concentrated sulfuric acid. The excess
ceric ion is then titrated with standard-ferrous ion using ferroin as
the indicator. The standard deviation is ± 0.17% [10].

Method 3. Hypophosphorous acid and phosphorous acid may be
determined in a mixture of the two by oxidation of the latter to orthophosphate by the addition of excess 0.1 N silver perchlorate. The
excess is back-titrated with 0.1 N sodium chloride. Samples containing
20-65 mg hypophosphite are analyzed and the end point is determined
potentiometrically. In a separate sample, the sum of the two acids is
determined by the addition of excess 0.1 N bromate-bromide solution,
the mixture is made 0.11 N with respect to hydrochloric acid, excess
potassium iodide is added and the liberated iodine is titrated in the
usual fashion with standard thiosulfate. Phosphorous acid may be determined by the procedure of Norkus et al. [9 and 11].

Precipitation Titrations of Orthophosphate

     Method 1. Samples containing 0.03-0.3 mole phosphate are titra-
ted potentiometrically with a standard silver ion solution in a borate
buffer in a pH range of 7.5-9 using a silver electrode versus silver-
silver chloride reference electrode. Although there is a slight deviation from the theoretical stoichiometry corresponding to the formation
of silver phosphate as the phosphate concentration increases, precision
and accuracy are of such a nature that the use of a linear calibration
curve is justified. This deviation is probably due to the adsorption
of phosphate ions on the precipitate of silver phosphate. Fluoride,
nitrate, sulfate, and acetate do not interfere. Chloride, bromide, or
iodide ion can be determined in addition to phosphate ion in a single
titration. All cations such as calcium, magnesium, aluminum, etc.
which precipitate at the pH of the titration interfere but may be easily removed by ion-exchange techniques. The relative standard deviation is 0.1% [12].

     Method 2. Orthophosphate is titrated with standard silver nit-
rate in a medium of 80% ethyl alcohol which is 0.1 M in sodium acetate.
The silver can also be obtained by coulometric generation from a silver
anode at a rate of 19.3 mg. All potential measurements are made with
a pH meter using a silver wire indicating electrode versus SCE. The
end point may also be determined amperometrically. The minimum orthophosphate concentrations are 2×10^{-4} M using a potentiometric end
point detection and 1.7×10^{-3} when the end point is determined amperometrically. Halide ions interfere as well as calcium, aluminum, and
iron(III). The relative error is $\pm$ 0.9% [13].

     Method 3. Orthophosphate is titrated in 0.5-0.7 N nitric acid
with a standard bismuth(III) solution using as a mixed indicator xylenol orange-methylene blue or methylthymol blue-methylene blue. By the
addition of chloroform to coalesce the precipitate, a very sharp end
point is obtained. The method is rapid; the precision and accuracy
are excellent [14].

     Method 4. Macro- and semimicro amounts of orthophosphate are
titrated in aqueous solution at 60-90° with standard cerium(III) solution. The titration is carried out in a buffered medium at pH 7.0-9.5
using Eriochrome Black T as the indicator. Accuracy at the 95% confidence limits is $\pm$ 0.38% for samples containing 8-26% phosphorus; the
corresponding precision is $\pm$ 0.24%. The method is rapid, however,
polyphosphates interfere [15].

     Method 5. Orthophosphate is determined by forming phosphomolyb-
dic acid in an aqueous solution free from ammonium salts followed by
precipitation of quinoline phosphomolybdate. The precipitate of quinoline phosphomolybdate is less soluble than the ammonium phosphomolybdate, and is relatively free from adsorbed or occluded impurities.
The quinoline phosphomolybdate is subsequently decomposed by treatment
with 0.5 N sodium hydroxide in excess followed by back-titration with
0.5 N hydrochloric acid using a mixed indicator composed of phenolphthalein and thymol blue. Calcium, magnesium, iron, aluminum, salts
of the alkali metals, citric acid, and citrates do not interfere.

Chromium may be present up to 18 times the phosphorus content and titanium up to 3.5 the phosphorus content without effect. The vanadium to phosphorus ratio must not exceed 0.2:1 [16,17].

Method 6. Samples containing 30 μg of phosphate are precipitated as phosphomolybdic acid which is separated from excess molybdate by extraction with isobutyl acetate. The phosphomolybdate is then back-titrated and degraded with 4 N ammonium hydroxide. The ammoniacal solution is made 2 N with respect to hydrochloric acid and the 12 molybdate ions accompanying each phosphate are reduced on a silver reductor column to molybdenum(V). The sample is eluted from the column with hot 2 N hydrochloric acid, and the molybdenum(V) is titrated with 0.001 M cerium(IV) sulfate using a ferroin indicator. Elements such as arsenic, antimony, germanium, and silicon which form heteropoly acids do not interfere. The end point may also be determined spectrophotometrically. The standard deviation for the titration of 929 μg of molybdenum (equivalent to 25 μg phosphorus) is 0.55%. The standard deviation for the whole procedure is 1.3% [18].

References

[1]. N. H. Furman, ed., *Standard Methods of Chemical Analysis*, 6th Ed., D. Van Nostrand, Princeton, N. J., 1962, pp. 816-818.

[2]. A. Schwicker, *Z. Anal. Chem.*, *110*, 161 (1937).

[3]. Boyer and Bauzil, *J. Pharm. Chim. (7)18*, 321 (1918).

[4]. I. M. Kolthoff, *Pharm. Weekblad*, *53*, 913 (1916).

[5]. G. G. Rao and H. S. Gowda, *Z. Anal. Chem.*, *146*, 167 (1955).

[6]. N. H. Furman, ed., *Standard Methods of Chemical Analysis*, 6th Ed., D. Van Nostrand, Princeton, N. J., 1962, p. 814.

[7]. A. F. Schmeckenbecher and J. A. Lindholm, *Anal. Chem.*, *39*(8), 1014-1016 (1967).

[8]. G. G. Guilbault and W. H. McCurdy, Jr., *Anal. Chim. Acta*, *24*, 214-218 (1961).

[9]. P. Norkus, A. M. Luneckas, and P. S. Carankute, *Zh. Analit. Khim.*, *20*(6), 753-755 (1965).

[10]. G. G. Guilbault and W. H. McCurdy, Jr., *Anal. Chim. Acta*, *24*, 214-218 (1961).

[11]. G. Ackermann and A. Mende, *Z. Anal. Chem.*, *232*(2), 97-103 (1967).

[12]. D. H. McColl and T. A. O'Donnell, *Anal. Chem.*, *36*(4), 848-850 (1964).

[13]. G. D. Christian, E. C. Knobloch, and W. C. Purdy, *Anal. Chem.*
       35(12), 1869-1871 (1963).

[14]. E. Bakács-Polgár, *Z. Anal. Chem., 190,* 373-376 (1962).

[15]. T. A. Taulli and R. R. Irani, *Anal. Chem.,* *35*(8), 1060-1063
       (1963).

[16]. H. N. Wilson, *Analyst, 76,* 65 (1951).

[17]. H. N. Wilson, *Analyst, 79,* 535 (1954).

[18]. G. S. Kirkbright, *Analyst, 93,* 224-227 (1968).

ARSENIC

The recommended methods for the determination of arsenic involve redox titration.

Synoptic Survey

Classical Methods

Redox Titration. Arsenic(III) may be determined by titration with standard iodine in an acid carbonate buffered solution [1], with standard bromate [2], with standard permanganate in the presence of iodate as a catalyst [3], with standard potassium iodate in hydrochloric acid [4], or with standard ceric sulfate using osmium tetroxide as the catalyst [5].

Precipitation Titration. Arsenate may be determined indirectly by precipitation as silver arsenate which is dissolved in nitric acid and the silver titrated by the Volhard thiocyanate method [6].

Contemporary Methods

Redox Titrations. Arsenite is titrated with standard hydrogen peroxide in a basic solution using lucigenin as the indicator [7].
    Arsenite is determined by titration in a sodium acetate buffer with standard iodine trichloride [8].
    Arsenite is titrated in the presence of arsenate with standard N-bromosuccinimide [9].
    Total arsenic(III) and arsenic(V) is determined, following a prior reduction step, with electrogenerated iodine. The end point is measured coulometrically [10].
    Arsenite is titrated potentiometrically with standard peroxymolybdate [11].

Precipitation Titration. Arsenate is determined indirectly by precipitation as silver arsenate which is dissolved and titrated potentiometrically with standard thiocyanate [12].

Outline of Recommended Contemporary Methods for Arsenic

Redox Titrations

Method 1. Arsenite is titrated with standard hydrogen peroxide in a basic solution using lucigenin (bis-N-methylacridinium nitrate) as the indicator. The end point is detected by the chemiluminescence which appears upon the addition of the slightest excess of hydrogen peroxide [7].

Method 2. Arsenite is titrated in a solution buffered with sodium acetate at a pH of 6.5-7.5. The titration is performed in the presence of $CHCl_3$ using 0.05 M iodine trichloride. The end point is determined by the appearance of the violet color of iodine in the chloroform layer. A potentiometric end point determination is also feasible [8].

Method 3. Arsenite in the presence of arsenate is titrated at room temperature with freshly prepared 0.1 N N-bromosuccinimide in a solution containing sodium bicarbonate and potassium iodide. During the titration, the titrant is irreversibly reduced to succinimide. The end point is indicated by a permanent blue color in the solution being titrated. As little as 100 µg of arsenic trioxide can be determined with an error of about 2% [9].

Method 4. Total arsenic and arsenic(III) are determined by the constant current anodic generation of iodine in neutral solution. Total arsenic is found by reducing the arsenic(V) to arsenic(III) with excess iodide in a strong acid solution containing borax. The iodine so formed is removed on a Dowex 3 anion exchange resin. The presence of the borax minimizes the elution of any antimony(III). The pH of the eluent is raised to pH 7, iodide is added, and the titration is carried out using a constant current of 6.43 mA. The end point is found amperometrically using a platinum indicating electrode polarized at -0.15 V. Readings are taken several times before and after the end point when the indicator electrode registers values between 5 and 15 µA. The straight-line plot of indicator current versus microequivalents is extrapolated to the residual current to establish the end point.
Arsenic(III) is found by using the same procedure but omitting the reduction with iodide prior to elution through the ion exchange column. The results agree favorably with other accepted methods for arsenic [10].

Method 5. Samples which are 0.01 N in arsenite are acidified with 1.5 N sulfuric acid and titrated potentiometrically with 0.02 N sodium peroxymolybdate. A few drops of 0.5 N potassium iodide added to the solution being titrated seems to catalyze the reaction. The titration is carried out using platinum reference electrode versus SCE [11].

References

[1]. N. H. Furman, ed., *Standard Methods of Chemical Analysis*, 6th Ed., Vol. 1, D. Van Nostrand, Princeton, N. J., 1962, pp. 114-115.

[2]. N. H. Furman and C. O. Miller, *J. Am. Chem. Soc.*, *59*, 152 (1937).

[3]. R. Lang, *Z. Anorg. Allgem. Chem.*, *152*, 197-206 (1926).

[4]. N. H. Furman, ed., *Standard Methods of Chemical Analysis*, 6th Ed., Vol. 1, D. Van Nostrand, Princeton, N. J., 1962, p. 116.

[5]. K. Gleu, *Z. Anal. Chem.*, *95*, 305-310 (1933).

[6]. L. W. McCay, *Am. Chem. J.*, *8*, 77 (1886).

[7]. L. Erdey and L. Buzás, *AD 627166*, *U.S. Gov't. Res. Develop. Rpt.*, *41*(5), 21 (1966).

[8]. B. Singh and G. P. Kashyap, *Indian J. Appl. Chem.*, *22*, 15-18 (1959).

[9]. M. Z. Barakat and A. Abadalla, *Analyst*, *85*, 288-294 (1960).

[10]. W. M. Wise and J. P. Williams, *Anal. Chem.*, *36*(1), 19-21 (1964).

[11]. S. Kotkowski and A. Lassocinska, *Chem. Anal. (Warsaw) 11*(4), 789-791 (1967).

[12]. V. Kuladaivelu and A. P. Madhaven, *Altech.*, *14*, 43-55 (1965).

ANTIMONY

The most common method for the determination of antimony is by redox titrimetry. An extensive review is available which covers the literature up to 1953 [1].

Synoptic Survey

Classical Methods

Redox Titrations. Antimony(III) is titrated with standard potassium bromate in hydrochloric acid solution [2], with standard sodium hypochlorite [3], with standard potassium permanganate [4], conductometrically or visually with standard ceric sulfate [5], or with standard iodine [6].

Antimony(V) is determined by reduction with potassium iodide in strong hydrochloric acid solution and the liberated iodine is titrated with standard thiosulfate [7].

Contemporary Methods

Redox Titrations. Antimony(III) is determined by titration with standard bromate using dead-stop end-point measurements [8].

Antimony(III) is determined by oxidation with standard potassium manganate in the presence of a telluric acid catalyst [9].

Antimony(III) is titrated with standard perbenzoic acid [10].

Antimony(V) is titrated potentiometrically with standard solutions of chromium(II).

Outline of Recommended Contemporary Methods for Antimony

Redox Titrations

Method 1. The classic titration method based on the oxidation of antimony(III) to antimony(V) with bromate has the disadvantage that the methyl orange indicator is destroyed at the first excess of bromate, or else fades during the course of the titration due to partial oxidation. This difficulty is overcome by using a dead-stop method for the detection of the end point. Samples containing 0.25-2.0 gm antimony are titrated with 0.1 N potassium bromate using a 0.2 V polarizing potential and a dead-stop end-point apparatus. Tin, lead, and arsenic do not interfere. Samples should not contain more than 10 mg copper or 750 mg iron. Samples containing 11.23% antimony show a standard deviation of 0.05%.

Method 2. Antimony(III) in concentrations as low as 1×10^{-3} M is determined by oxidation with potassium manganate. To a known volume of standard manganate solution containing a small amount of telluric acid, there is added as the titrant the unknown antimony sample. The end point is determined potentiometrically. During the titration, the following redox reaction occurs:

$$Sb(III) \quad + \quad Mn(VI) \quad \rightarrow \quad Sb(V) \quad + \quad Mn(IV)$$

Low results are obtained if the manganate solution is added to the unknown antimony sample due to further reduction of the manganese. Precision and accuracy are very good [9].

Method 3. Antimony(III) in samples such as tartar emetic is titrated with 0.05 M perbenzoic acid in chloroform in the presence of iodine as a catalyst and preoxidizer. The end point is marked by a faint violet color in the chloroform layer. The end point may also be determined potentiometrically using a platinum foil electrode versus SCE. Precision and accuracy are excellent [10].

Method 4. Samples containing 21-200 mg antimony(V) in 30% hydrochloric acid are titrated at room temperature with 0.05-1.5 N chromium(II). The end point is determined potentiometrically using a platinum wire electrode versus SCE. Tin, arsenic, zinc, aluminum, cadmium, and small amounts of lead do not interfere. Interferences are noted from iron, copper, and tellurium. The error is less than 0.4%. The titration of antimony(V) may also be carried out using vanadium(II) as the titrant at room temperature. In this case, tin, arsenic, zinc, aluminum, cadmium, beryllium, tellurium, and small amounts of lead do not interfere; iron and copper do. In the titration with vanadium(II) the error is $\leq$ 0.7% [11].

References

[1].   D. Gibbons, *Ind. Chemist, 29,* 363, 418 (1953).

[2].   C. L. Luke, *Ind. Eng. Chem. Anal. Ed., 16,* 448 (1944).

[3].   N. I. Goldstone and M. B. Jacobs, *Ind. Eng. Chem. Anal. Ed., 16,* 206 (1944).

[4].   W. Pugh, *J. Chem. Soc., 1933,* 2-4.

[5].   H. H. Willard and P. Young, *J. Am. Chem. Soc., 50,* 1372 (1928).

[6].   N. H. Furman and C. O. Miller, *J. Am. Chem. Soc., 59,* 152 (1937).

[7].   N. H. Furman, ed., *Standard Methods of Chemical Analysis,* 6th Ed., Vol. 1, D. Van Nostrand, Princeton, N. J., 1962, pp. 96-97.

[8].   G. Bradshaw, *Analyst, 88*(1049), 599-602 (1963).

[9].   G. den Boef and A. Daalder, *Z. Anal. Chem., 167,* 430-431 (1959).

[10]. B. Singh, S. S. Sahota, and A. Singh, *Z. Anal. Chem., 169,* 106-109 (1959).

[11]. L. Naruskevicius, *Lietuvos TSR Aukstuju Mokyklu Mokslo Darbai, Chem. ir Chem. Technol., 7,* 149-156 (1965).

BISMUTH

Methods for the determination of bismuth include redox, complexometric, and precipitation titrations.

Synoptic Survey

Classical Methods

Redox Titrations. Bismuth is precipitated as the oxalate which is hydrolyzed to the basic oxalate and titrated with standard permanganate in the presence of sulfuric acid [1,2].

Bismuth is titrated with standard EDTA at pH 1.5-2.0 in the presence of thiourea as a complexer and indicator. A photometric or potentiometric determination of the end point may also be used [3,4,5].

Contemporary Methods

Redox Titrations. Bismuth is determined by reduction with glucose
or fructose to the metal which is added to standard iron(III) and the
iron(II) formed is titrated with standard cerium(IV) solution [6].

Bismuth solutions are treated with metallic copper foils, and
the copper(I) so formed is titrated with anthranyldiacetic acid using
murexide indicator [7].

Complexometric Titrations. Bismuth is precipitated with bismuthiol (2,5-dimercapto-1,3,4-thiadiazole). The precipitate is added to
an excess of standard EDTA and the excess is back-titrated with standard magnesium sulfate [8].

Bismuth is titrated amperometrically with standard 3,5-diethyl-
2,6-dimercapto-1,4-thiopyrone [9].

Bismuth is titrated spectrophotometrically with standard EDTA
using an iron-salicylic acid complex as the indicator [10].

Precipitation Titrations. Bismuth is titrated potentiometically
with standard solutions of ferrocyanide [11]. The end point may also
be determined visually using variamine blue as the indicator [12].

Bismuth is precipitated as the benzene seleninate or the naphthalene seleninate. The precipitate is dissolved, excess potassium
iodide is added and the free iodine is titrated with standard thiosulfate [13].

Bismuth is titrated with standard potassium dihydrogen phosphate
in dilute nitric acid using xylenol orange as the indicator [14].

Outline of Recommended Contemporary Methods for Bismuth

Redox Titrations

Method 1. Samples of bismuth solutions are boiled 3-4 minutes
with excess D-glucose or D-fructose. The bismuth formed during the
reduction is filtered and reoxidized by boiling for 15-20 minutes in
4 N hydrochloric acid with excess standard iron(III) chloride. The
iron(II) formed is then titrated with standard cerium(IV) solutions
in 8 N sulfuric acid using N-phenylanthranilic acid as the indicator.
Aluminum and thorium do not interfere. Interferences are noted from
lithium, barium, copper, strontium, magnesium, cadmium, zinc, lead,
nickel, cobalt, mercury, gold, selenium, zirconium, and tellurium.
The method is well-suited to the determination of microquantities of
bismuth [6].

Method 2. Copper foils are added to samples of bismuth containing 15-30 mg in hydrochloric acid. After heating for about 15 minutes
at the boiling point, the solutions are decanted, and the copper(I)
is titrated at pH 7.5 with standard anthranyldiacetic acid using murexide as the indicator. Ions which interfere with the complexometric
titration of copper(I) or which are reduced by metallic copper must
be absent; chloride does not interfere. The error is $\pm$ 1.0% [7].

Complexometric Titrations

Method 1. Bismuth is precipitated from 0.1 N hydrochloric acid or sulfuric acid with bismuthiol. After filtering, and washing, the precipitate is dissolved in a known excess of 0.02 M EDTA. The pH is adjusted with an ammonium hydroxide-ammonium chloride buffer, and the excess EDTA is back-titrated with standard magnesium sulfate to an Eriochrome Black T end point. By this procedure involving Bismuthiol, bismuth is separated from iron(II), aluminum, chromium, cerium(III), zirconium, titanium, zinc, thorium, uranyl ion, beryllium, mganesium, manganese, cobalt, nickel, alkali metals, alkaline earths, and rare earths. Mercury(II), lead, copper(II), silver, thallium(I), cadmium, and palladium interfere but may be removed by precipitation with the Bismuthiol at a pH 6.0-8.5 in the presence of tartaric acid and ammonium chloride. Bismuth is then precipitated as above after adjustment of the pH to 1.4-2.5. The following ions interfere with the titration: antimony(III), tin(II), iron(III), fluoride, vanadate, phosphate, arsenate, and chromate. The precision and accuracy are very good.

Method 2. Samples containing 1.6-20 mg bismuth are titrated amperometrically at pH 4.27 in an acetate buffer with 0.1 N 3,5-diethyl-2,6-dimercapto-1,4-thiopyrone. The electrode system consists of a waxed graphite electrode versus SCE. The applied potential is + 0.5 V and the end point is determined graphically. The relative error for samples containing 1.6-20 mg/25 ml bismuth is $\leq$ 2% [9].

Method 3. Bismuth is titrated spectrophotometrically with standard EDTA using an iron-salicylate complex as the indicator. The titration is carried out at pH 0.5 in a solution buffered with a sodium acetate-nitric acid buffer. The absorbance is measured at 520 mμ. Although iron(III) is usually a serious interference in the titration of bismuth, due to the differences in the stability constants of the metal-EDTA complexes involved, iron(III) does not react with the bismuth-EDTA complex. From 8-500 mg bismuth per liter can be determined; the error is $\leq$ 1% [10].

Precipitation Titrations

Method 1. Bismuth is titrated with standard solutions of alkali metal or ammonium ferrocyanides at pH 2-6. The titration is carried out using a platinum foil electrode versus SCE and the potential change at the end point is enhanced by the addition of 20% ethyl alcohol and ammonium nitrate. Best results are obtained using ammonium ferrocyanide [11]. The end point may also be determined visually or spectrophotometrically using variamine blue as the indicator. Accuracy and precision are excellent and low concentrations of bismuth can be determined. Ions oxidizing or reducing the indicator or those which form stable compounds with bismuth must be absent [12].

Method 2. Bismuth is precipitated with ammonium benzene seleninate or ammonium naphthalene seleninate yielding $Bi(C_6H_5SeO_2)_3$ or $Bi(C_{10}H_7SeO_2)_3$. The precipitate is dissolved in hydrochloric acid and tartaric acid; excess potassium iodide is added and the liberated iodine is titrated with standard sodium thiosulfate. The relative error for 5-20 mg bismuth is less than 0.5% [13].

Method 3. Bismuth is titrated with standard potassium dihydrogen phosphate in 0.5-0.7 N nitric acid using a mixed indicator composed of xylenol orange and methylene blue or methylthymol blue and methylene blue. An equal volume of chloroform is added prior to the titration with thiosulfate to coalesce the precipitate and sharpen the end point. The method is rapid; precision and accuracy are very good [14].

References

[1]. O. Warniak and O. Kyle, *Chem. News, 75*, 3 (1897).

[2]. M. M. Muir and C. E. Robbs, *J. Chem. Soc.*, *41*, 1 (1882). see also: C. L. Wilson and D. W. Wilson, *Comprehensive Analytical Chemistry*, Elsevier, N. Y., Vol. I, 1962, pp. 268-270.

[3]. J. S. Fritz, *Anal. Chem., 26*, 1978 (1954).

[4]. A. L. Underwood, *Anal. Chem.*, *26*, 1322 (1954).

[5]. R. N. White and A. L. Underwood, *Anal. Chem.*, *27*, 1334 (1955).

[6]. O. G. Soxena, *Chim. Anal.* (Paris), *49*(7), 384-385 (1967).

[7]. C. Dragulescu, D. Lazar-Jucu, and R. Kuzman-Anton, *Bul. Stiint. Teh. Inst. Politeh. Timisoara.*, *11*(1), 69-71 (1966).

[8]. A. K. Majumdar and M. M. Chakrabartty, *Z. Anal. Chem.*, *165*, 100-105 (1959).

[9]. A. M. Arishkevich, A. S. Akhmetshim, and Yu. I. Usatenko, *Zvod. Lab.*, *33*(6), 692-695 (1967).

[10]. N. A. Ramaiah, G. D. Tewari, S. R. Trivedi, and S. S. Katiyar, *Talanta*, *15*(3), 352-355 (1968).

[11]. R. S. Saxena and C. S. Bhatnagar, *Z. Anal. Chem.*, *165*, 94-99 (1959).

[12]. A. Gregorowicz and B. Piwowarska, *Mikrochim. Ichnoanal. Acta*, *1963*(4), 755-758.

[13]. V. S. Sotnikov and I. P. Alimarin, *Zhur. Anal. Khim.*, *14*, 710-713 (1959).

[14]. E. Bakács-Polgár, *Z. Anal. Chem.*, *190*, 373-376 (1962).

VANADIUM(II) AND VANADIUM(III)

Methods for the determination of vanadium(II) and vanadium(III) involve redox and complexometric titrations.

Synoptic Survey

Classical Methods

None of importance.

Contemporary Methods

Redox Titrations. Vanadium(II) is oxidized with excess iron(III) which is then back-titrated with standard chromium(II) [1]. Vanadium(II) and vanadium(III) may be determined in the same sample by a double titration. In one aliquot, the vanadium(II) is air oxidized to vanadium(III) which is titrated with standard ammonium metavanadate. In a second aliquot, the vanadium(II) and vanadium(III) are oxidized with iron(III) to vanadium(IV) and again titrated with standard metavanadate [1].

Complexometric Titration. Vanadium(III) is titrated amperometrically with standard iron at pH 4 [2].

Outline of Recommended Contemporary Methods for
Vanadium(II) and Vanadium(III)

Redox Titrations

Method 1. Vanadium(II) is oxidized with excess standard iron (III) solution in the presence of orthophosphoric acid. If vanadium (III) is present, only the vanadium(II) is oxidized. The excess iron (III) is back-titrated potentiometrically with standard chromous solution using a tungsten indicator electrode. The accuracy varies between 0.2 and 1% [1].

Method 2. Vanadium(II) and vanadium(III) may be determined individually in a mixture by double titration. In one aliquot, the vanadium(II) is oxidized to vanadium(III), and the total trivalent vanadium is determined by titration with 0.1 N ammonium metavanadate using sodium diphenylamine sulfonate as the indicator. In a second aliquot, vanadium(II) and vanadium(III) are oxidized by the addition of excess standard ferric ammonium sulfate to vanadium(IV). Iron(II) is formed in an amount equivalent to vanadium(II) and vanadium(III). Titration is carried out using 0.1 N ammonium metavanadate. The error is 0.2%.

Complexometric Titration

Vanadium(III) in the presence of vanadium(IV) is titrated amperometrically at 1.1 V with standard tiron at pH 4.0 in an acetate buffer. Vanadium(III) forms a 1:1 complex with tiron. The instability constant of the complex is about 1×10^{-4}. The end point may also be determined spectrophotometrically at 390 mµ [2].

References

[1]. L. I. Veselago, *Zh. Analit. Khim.*, *20*(3), 335-338 (1965); English transl. *J. Anal. Chem. USSR*, *20*(3), 298-311 (1965).

[2]. I. A. Tserkovnitskaya and M. F. Grigoreva, *Zh. Analit. Khim.*, *21*(11), 1395-1398 (1966); English transl. *J. Anal. Chem. USSR*, *21*(11), 1240-1244 (1966).

VANADIUM(IV)

Methods for the determination of vanadium(IV) involve redox and complexometric titrations.

Synoptic Survey

Classical Methods

Redox Titration. Vanadyl ion is titrated with standard solutions of permanganate [1].

Contemporary Methods

Redox Titrations. The standard method for the titration of vanadium(IV) with permanganate is improved by the addition of orthophosphoric acid as a catalyst and the use of a ferroin indicator [2].
Vanadium(IV) is titrated photometrically with 0.2 N potassium dichromate in orthophosphoric acid at 660 mμ [3]. The titration may also be performed with standard ceric sulfate with ferroin indicator [4].

Complexometric Titrations. Vanadium is titrated at pH 3 with 0.01 M EDTA and N-benzoyl-N-phenylhydroxylamine as the indicator [5].
Vanadium(III) and vanadium(IV) may be titrated in succession with 0.1 N trilon-B [6].

Outline of Recommended Contemporary Methods for Vanadium(IV)

Redox Titrations

Method 1. Vanadium(IV) is titrated to vanadium(V) at room temperature with standard potassium permanganate in 0.2-0.3 N sulfuric acid containing orthophosphoric acid as a catalyst. No indicator is required with the 0.1 N permanganate, but a blank should be run. If the titration is carried out with 0.1 N potassium permanganate, ferroin is used as the indicator, and a blank is not needed. Colored ions such as chromium(III) up to a reasonable limit do not interfere. Precision and accuracy are excellent [2].

Method 2. Samples containing 10-50 mg vanadium(IV) are titrated photometrically with 0.2 N potassium dichromate in 3 M orthophosphoric acid. The end point is found at 660 mμ. The absorbance decreases as vanadium(IV) is oxidized to vanadium(V) and becomes constant at the end point. The maximum relative error is 0.7%. Iron(II), arsenic(III), antimony(III), uranium(IV), and molybdenum(V) are oxidized by the potassium dichromate and interfere. In the case of iron(II) a separate equivalence point can be obtained from the photometric titration curve. Iron(III), cerium(III), manganese(II), cobalt(II), nickel(II), molybdenum(VI), uranium(VI), chromium(III), zinc(II), chloride, and nitrate do not interfere. Tungsten(VI) interferes, but the interference can be eliminated by running the titration in 7.5 M or higher concentrations of orthophosphoric acid. The photometric method has the advantage over the potentiometric titration in that chloride, nitrate, cerium (III), and manganese(II) do not interfere. Precision and accuracy are good [3].

Method 3. Vanadium is titrated with standard ceric sulfate in acid solution using ferroin as the indicator. This method may also be employed in the titration of a mixture of iron(II) and vanadium(IV) provided that the sulfuric acid concentration is 5 M at the equivalence point. At this concentration of acid, the formal redox potential of ferroin is lowered (0.925 V) and that of vanadium(V) is increased (1.190 V) so that the indicator is oxidized preferentially. As much as 55 mg of iron(III), 9 mg of chromium(III) and 84 mg of vanadium(V) do not interfere. Precision and accuracy are very good [4]. The titration may also be carried out in 0.5-1 M sulfuric acid using methyl orange as the indicator [5].

Complexometric Titrations

Method 1. Vanadium(IV) is titrated at pH 2.5-4.5 with 0.01 M EDTA and N-benzoyl-N-phenylhydroxylamine as the indicator. At the end point the color change is from reddish to a sky blue. Since the vanadium-benzoylphenylhydroxylamine-complex is unstable at higher temperatures, the titration must be performed at room temperature. Samples containing vanadate are reduced to vanadyl with sulfite prior to titration. Samples containing 9.6 mg of vanadium may be analyzed in the presence of 23.5 mg of titanium(IV) or 10 mg of molybdenum(VI) without interference. Vanadium(IV) may also be determined in the presence of manganese(II). Amounts up to 150 mg of potassium fluoride and 200 mg of diammonium hydrogen phosphate do not interfere in the determination of 7.5-15 mg of vanadium. Iron(III) must be removed by precipitation prior to reduction of vanadate to vanadyl and subsequent titration. Precision and accuracy are very good. The error is less than 3 parts per thousand [6].

Method 2. Vanadium(III) and vanadium(IV) may be titrated separately or in a mixture using the tetrasodium salt of EDTA (Trilon B) as the titrant. The end point is determined by anode amperometric titrimetry. Vanadium(III) and vanadium(IV) form complexes with trilon B which are quite different in stability. The dissociation constant of the vanadium(III) complex is one-seventieth that of the VO^{+2}-complex. If the sample contains only vanadium(III) or vanadium(IV), it is titrated at pH 1.0-4.0 with 0.1 N trilon B. If both are present, the titration is carried out at pH 1.0-2.5 in sulfuric acid solution. At this pH the titration curve shows only one break due to the vanadium (III). If the titration is repeated at pH 3-4, also in sulfuric acid solution, two breaks in the titration curve are observed. The first one indicating the amount of vanadium(III) and the second one corresponding to the amount of vanadium(IV) [7].

References

[1]. N. H. Furman, ed., *Standard Methods of Chemical Analysis*, 6th Ed., Vol. 1, D. Van Nostrand, Princeton, N. J., 1962, pp. 1211-1212.

[2]. L. S. A. Dikshitulu and G. G. Rao, *Z. Anal. Chem.*, *189*, 421-426 (1962).

[3]. G. G. Rao and P. K. Rao, *Talanta*, *14*(1), 33-43 (1967).

[4]. K. Sriramam and G. G. Rao, *Talanta*, *13*(10), 1468-1469 (1966).

[5]. K. R. Rao, *Chem.-Analyst*, *54*(4), 104-105 (1965).

[6]. V. R. M. Kaimal and S. C. Shome, *Anal. Chim. Acta*, *27*, 594-596 (1962).

[7]. I. A. Tserkovnitskaya and N. A. Kustova, *Vestnik Leningrad. Univ. 15, No. 16, Ser. Fiz. i. Khim. No. 3*, 148-149 (1960).

VANADIUM(V)

Methods for the determination of vanadium(V) include redox, complexometric and precipitation titration.

Synoptic Survey

Classical Methods

Redox Titrations. Vanadium(V) is reduced with sulfur dioxide or hydrogen sulfide and the resulting vanadium(IV) is titrated with standard potassium permanganate [1].

Vanadium(V) is reduced to vanadium(II) with zinc in sulfuric acid. Re-oxidation with iron(III) yields iron(II) which is titrated with standard permanganate [2].

Vanadium(V) is titrated with standard ferrous sulfate either visually or potentiometrically [3,4].

Contemporary Methods

Redox Titrations. Vanadium(V) is titrated coulometrically with electrogenerated tin(II). The end point is determined spectrophotometrically or potentiometrically [5,6]. The titration may also be carried out with electrogenerated uranium(V) and the end point detected amperometrically [7].

Vanadium(IV) and vanadium(V) are determined in a mixture by a double titration involving Mohr's salt for the first, and permanganate for the second subsequent titration [8].

Vanadium(V) is titrated with standard thiosulfate in the presence of cupric ion as a catalyst [9].

Complexometric Titrations. Vanadium(V) is titrated photometrically with standard pyridine-2,6-dicarboxylic acid in the presence of H_2O_2 [10].

Vanadium(V) is titrated with standard EDTA in the presence of ferrous ion using variamine blue as the indicator [11].

Vanadium(V) is reduced to vanadium(IV); excess standard EDTA is added and the excess is back-titrated with standard mercuric ion solution. The end point is detected potentiometrically [12].

Vanadium(V) is titrated potentiometrically with mercury-EDTA solutions [13].

Precipitation Titration. Vanadium(V) is titrated amperometrically with standard silver nitrate at pH 8-9 [14,15].

Outline of Recommended Contemporary Methods for Vanadium(V)

Redox Titrations

Method 1. Vanadium(V) is titrated coulometrically to vanadium (IV) by electrogenerated tin(II). The supporting electrolyte consists of 4 M sodium bromide, 0.2 M stannic chloride, and 0.25 N hydrochloric acid. With this electrolyte, the current efficiency for the reduction of tin(IV) is 99.5-99.9% and generating current densities of 10-84 mA/cm^2 at a gold generator cathode are used. The end point is detected potentiometrically or spectrophotometrically by measuring the change in absorbance at 390 mμ. At this wavelength vanadium(V) and tin(II) both absorb strongly, however, vanadium(IV) and tin(IV) do not. The titration curve is V-shaped with the end point at minimum absorbance. The formal potential of the V(V)/V(IV) couple is + 0.63 V versus SCE which is high enough to cause the reduction of some vanadium(V) by bromide ion during the titration. It is, therefore, necessary to add some vanadium(IV) to the supporting electrolyte to keep the potential of the V^{+5}/V^{+4} couple below the critical value at which bromide ion is oxidized. Hence, the supporting electrolyte containing the vanadium(IV) must be pretitrated to the end point before the sample of vanadium(V) is added. For samples containing 0.5 mg and 5.5 mg vanadium, the average deviation is $\pm$ 0.8% and $\pm$ 0.2% respectively [5,6].

Method 2. Vanadium(V) is titrated coulometrically with electrogenerated uranium(V) from uranyl chloride at pH 1.5-1.9. The end point is obtained amperometrically. Iron(III) is reduced simultaneously with vanadium(V). The error is 0.10% in the titration of 0.01-2.5 mg vanadium [7].

Method 3. In a mixture of vanadium(IV) and vanadium(V), the constituents are titrated amperometrically in 2 N sulfuric acid solution with two platinum indicator electrodes at 0.4 V. A double titration is performed. The first titrant is 0.1 N Mohr's salt which is added until a sudden increase of current indicates excess iron(II) and the formation of the redox couple Fe^{+3}/Fe^{+2}. The titration is then continued with the second titrant which is 0.1 N potassium permanganate. The abrupt drop in current is due to the oxidation of excess Fe^{+2}. The current then remains constant, and eventually increases sharply after the end point due to the cathodic reduction of excess permanganate ion. The amount of Mohr's salt used gives the vanadium(V) value. The amount of vanadium(IV) is equal to the difference of potassium permanganate minus Mohr's salt minus the potassium permanganate needed for the oxidation of the iron(II) ion. Relative error is less than 1% for vanadium(V) and about 1.5% for vanadium(IV) [8].

Method 4. Although the reaction of vanadium(V) with the thiosulfate ion is slow, it may be adapted to a direct titrimetric method by the use of iodine monochloride or copper(II) ion as a catalyst. Samples containing up to 1.25 mg vanadium(V) in 60 ml of a hydrochloric acid-sodium acetate buffer at pH 1.09 are treated with a few milliliters of 0.1 M copper(II) sulfate, diluted to 100 ml and titrated potentiometrically with 0.05 M sodium thiosulfate using a platinum wire electrode versus SCE. Vanadium(V) is reduced to vanadium(IV). Mixtures containing iron up to 10 times the concentration of vanadium can be titrated. In this case the titration curve shows two breaks. The first which is small corresponds to the reduction of vanadium(V) while the second corresponds to the reduction of iron(III). There are no appreciable interferences from iron, manganese, chromium, molybdenum, uranium, perchlorate, nitrate, oxalate, phosphate, or fluoride. The error is less than 1% [9].

Complexometric Titrations

Method 1. Vanadium(V) forms in mineral acid solution containing hydrochloric and phosphoric acids with 30% hydrogen peroxide and pyridine-2,6-dicarboxylic acid a 1:1:1 complex. The latter is used as the titrant in the photometric titration of samples of vanadium(V) in aqueous solution at 432-6 mµ. Vanadium can thus be titrated in the presence of iron and other alloying elements in steel. No separations or masking procedures are required. The error for 0.1-2.0 mg vanadium is approximately ± 2% [10].

Method 2. Samples containing 1×10^{-3} moles vanadium(V) are acidified to a pH 1.7-2.0 with sulfuric acid. A few milliliters of 0.1 M Mohr's salt is added, and the vanadium is titrated visually with 0.1 M EDTA using variamine blue as the indicator. The error is less than 0.5% [11].

Method 3. Samples of vanadium(V) are reduced in sulfuric acid solution with sodium sulfite. The mixture is then boiled to remove sulfur dioxide. Excess 0.05 M EDTA is added, and after buffering the solution at pH 8 with urotropine, the excess EDTA is back-titrated potentiometrically with 0.05 M mercuric ion. A silver amalgam electrode versus SCE is used for detection of the end point. The percentage of error is less than 0.7% [12].

Method 4. Vanadium(V) is titrated potentiometrically with standard Hg-EDTA solutions in an ammonium acetate-acetic acid buffer at pH 5.6 using an amalgamated gold thimble as the indicator electrode versus SCE as reference. For samples containing 5.095 mg vanadium, the percent deviation is $\pm$ 0.41% [13].

Precipitation Titration

Vanadium is titrated amperometrically with standard silver nitrate in a pH range of 8-9 at E_{dc} = -0.3 V versus SCE using a silver indicator electrode. As little as 0.5 mmole of sodium orthovanadate can be titrated with a precision of about $\pm$ 1% [14,15].

References

[1]. N. H. Furman, ed., *Standard Methods of Chemical Analysis,* 6th Ed., Vol. 1, D. Van Nostrand, Princeton, N. J., 1962, pp. 1211-1212.

[2]. N. H. Furman, ed., *Standard Methods of Chemical Analysis,* 6th Ed., Vol. 1, D. Van Nostrand, Princeton, N. J., 1962, p. 1213.

[3]. N. H. Furman, *Ind. Eng. Chem., 17,* 314 (1925).

[4]. G. L. Kelley, J. R. Adams, and J. A. Wiley, *J. Ind. Eng. Chem., 9,* 780 (1917).

[5]. A. J. Bard and J. J. Lingane, *Anal. Chim. Acta, 20,* 581-587 (1959).

[6]. A. J. Bard and J. J. Lingane, *Anal. Chim. Acta, 20,* 463 (1959).

[7]. S. L. Phillips and D. M. Kern, *Anal. Chim. Acta, 20,* 295-298 (1959).

[8]. O. A. Sogina and I. S. Savitskaya, *Zavodsk. Lab., 29,* 401-402 (1963).

[9]. V. P. R. Rao and B. V. S. Sarma, *Chem.-Analyst, 54*(4), 107-109 (1965).

[10]. H. Hartkamp, *Z. Anal. Chem., 171,* 272-280 (1959).

[11]. M. Tanaka and A. Ishida, *Anal. Chim. Acta, 36*(4), 515-521 (1966).

[12]. H. Khalifa and A. El-Siravy, *Z. Anal. Chem., 227*(2), 109-115 (1967).

[13]. M. C. Cardels and J. C. Cornwell, *Anal. Chem.*, *38*(6), 774-776 (1966).

[14]. R. S. Saxena and O. P. Sharma, *Indian J. Chem.*, *2*(12), 502-503 (1964).

[15]. R. S. Saxena and O. P. Sharma, *Talanta*, *11*(5), 863-866 (1964).

NIOBIUM

Methods for the determination of niobium include redox, complexation, and precipitation titration.

Synoptic Survey

Classical Methods

Redox Titration. Niobium(V) is reduced with amalgamated zinc to niobium(III) which is then titrated with standard potassium permanganate [1,2].

Contemporary Methods

Redox Titration. Niobium(V) is reduced to niobium(III), treated with excess ferric ammonium sulfate, and the ferrous ions formed are titrated with standard potassium dichromate [3].

Complexometric Titration. Niobium(V) is titrated in sulfuric acid solution with standard nitrilotriacetic acid (NTA) in the presence of hydrogen peroxide and using xylenol orange as the indicator [4,6]. Samples of niobium(V) containing hydrogen peroxide are treated with excess nitrilotriacetic acid, NTA, and the excess is back-titrated with standard copper solution [5,6].

Precipitation Titration. Niobium is titrated amperometrically with either cupferron or neocupferron [7].

Outline of Recommended Contemporary Methods for Niobium

Redox Titration

Samples containing niobium(V) are reduced in a mixture of 6 M hydrochloric acid and 0.5 M hydrofluoric acid in a Jones reductor charged with amalgamated zinc to niobium(III). The niobium(III) is collected in an excess of 0.04 N ferric ammonium sulfate acidified with orthophosphoric acid. Iron(II) so formed is titrated with 0.1 N potassium dichromate. From 1-50 mg of niobium may be titrated in the presence of different amounts of tantalum. The method is of

considerable importance in the determination of niobium in alloys.
Precision and accuracy are very good [3].

Complexometric Titrations

Method 1. Niobium in sulfuric acid solution is titrated at pH
1-3 with 0.05 M nitrilotriacetic acid, NTA, in the presence of a small
amount of 30% hydrogen peroxide. A stable ternary complex niobium-
NTA-H_2O_2 is formed; xylenol orange is used as the indicator. From 0.5
to 20 mg of niobium may be determined in a volume of 30-50 ml; titanium,
zirconium, vanadium, and tantalum interfere. Up to 4 mg tantalum can
be tolerated in samples of niobium of the above sizes. This is due
primarily to the faster rate of reaction of hydrogen peroxide with
niobium when compared to the rate of reaction with tantalum. The
method is recommended for use in the analysis of refractory compounds
[4,6].

Method 2. Samples containing niobium and hydrogen peroxide are
treated with an excess of a standard solution of NTA. The excess is
then back-titrated with standard copper solution at pH 5.0-5.5 using
the metalofluorescent indicator methylcalcein under ultraviolet illum-
ination. The complex formed has a niobium : H_2O_2 : NTA ratio of 1:1:1.
Since it is formed somewhat slowly, it is recommended that the NTA be
added to the hot solution of the sample at pH 3. Adjustment of the pH
to 5.0-5.5 is then made and the titration continued as above. The
reproducibility for 4.56-23.68 mg niobium is 0.07 mg [5,6].

Precipitation Titration

Niobium is titrated amperometrically with either standard cupferron or standard neocupferron. A platinum or a graphite electrode
is used versus SCE and the applied EMF is + 1.0 V. In the cupferron
titration the supporting electrolyte to be used is 2-18 N sulfuric
acid with sodium chloride as a coagulant. With neocupferron, 1-2 N
hydrochloric acid is used as the electrolyte. Neocupferron is a preferred titrant because of its greater selectivity and operability
over a wider concentration range. Tantalum does not interfere in the
titration. The error is less than 1% [7].

References

[1]. W. R. Schoeller and E. F. Waterhouse, *Analyst, 49*, 215 (1924).

[2]. H. B. Knowles and G. E. F. Lundell, *J. Res. Natl. Bur. Std. 42*,
405 (1949).

[3]. J. B. Headridge and M. S. Taylor, *Analyst, 87*, 43-48 (1962).

[4]. A. F. Kuteinikov and S. A. Lysenko, *Zavodsk. Lab., 33*(2), 141-145
(1967).

[5]. E. Lassner, *Talanta, 10*(22), 1229-1233 (1963).

[6]. A. F. Kuteinikov and S. A. Lysenko, *Zh. Analit. Khim.*, *22*(9)
      1366-1370 (1967).

[7]. Yu. I. Usatenko and A. P. Tikhonova, *Zavodsk. Lab.*, *33*(8) 939-942
      (1967).

TANTALUM

Close similarities in the chemical properties of tantalum and
niobium cause difficulties in the separation and determination of the
two elements from each other. A contemporary method for the determination of tantalum in the presence of niobium involves redox titration.

Synoptic Survey

Classical Methods

None of importance.

Contemporary Methods

Redox Titration. Tantalum is complexed by the addition of excess
hydrogen peroxide which is then back-titrated with standard potassium
permanganate [1].

Outline of a Recommended Contemporary Method for Tantalum

Redox Titration

Samples containing 0.001-0.002 mole of tantalum fluoride in 10-
20 ml solution are treated with 5-10 ml 1:1 sulfuric acid, 10-20 ml
phosphoric acid, 30-50 ml of a saturated solution of aluminum sulfate,
and 10-20 ml of 0.2 M hydrogen peroxide. The excess peroxide is then
titrated at 20° with 0.1 N potassium permanganate. The tantalum-
peroxy-complex is only slowly oxidized by permanganate at the above
temperature and at the concentrations of sulfuric acid and phosphoric
acid which are used. Niobium, if present, also forms a niobium-$H_2O_2^-$
complex which is rapidly destroyed by the phosphoric acid-sulfuric acid
mixture. Such behavior forms the basis for the determination of tantalum in the presence of niobium. One mole of peroxide is bound by
one mole of tantalum in 0.5-2 M sulfuric acid. At this concentration
of acid, the rate of reaction between bound peroxide and permanganate
is so slow that free hydrogen peroxide can be oxidized by the permanganate. In the presence of phosphoric acid in amounts 500 times that
of hydrogen peroxide, the niobium-peroxide complex reacts almost in-
stantly with the permanganate whereas the tantalum-peroxide complex
does not react in the time required for the titration. Thus the
amount of peroxide combined with the tantalum alone may be calculated
on the basis of the ratio $Ta:H_2O_2 = 1:1$. Titanium, niobium, molybdenum, vanadium, iron, and indium do not interfere [1].

References

[1]. A. K. Babko and I. G. Lukianets, *Zn. Analit. Khim.*, *21*(12),
      1430-1435 (1966). English transl., *J. Anal. Chem. USSR*, *21*(12),
      1273-1277 (1966).

GROUP VI

Oxygen	Polonium
Sulfur	Chromium
Selenium	Molybdenum
Tellurium	Tungsten

OXYGEN

In this section titrimetric procedures for elemental oxygen, water, hydrogen peroxide, and peroxo compounds will be discussed.

Synoptic Survey of Elemental Oxygen Determinations

The usual procedures for the determination of elemental oxygen involve redox titration.

Classical Methods

Redox Titration. Elemental oxygen oxidizes manganous hydroxide to manganic hydroxide which is used to liberate an equivalent amount of iodine from potassium iodide. Finally, the iodine is titrated with standard thiosulfate (Winkler's method) [1,2,3].

Contemporary Methods

Redox Titrations. Samples of oxygen are used to oxidize titanium(III) to titanium(IV). The excess titanium is then titrated with standard ferric chloride [4].

Dissolved oxygen is titrated coulometrically with electrogenerated chromous ion [5].

Samples of oxygen are reduced with hydrogen, and the formed water is titrated with the Karl Fischer reagent [6].

Samples of oxygen are used to oxidize manganese(II) to manganese(III). The manganese(III) is then complexed in acid solution with pyrophosphate and titrated with standard hydroquinone [7,8,9].

Outline of Recommended Contemporary Methods for Elemental Oxygen

Redox Titrations

Method 1. Gaseous samples containing 0.1-1.0% oxygen are bubbled through a basic solution of titanium(III) chloride containing calcium hydroxide. Titanium(III) is oxidized to titanium(IV) and precipitates as the hydroxide. The solution is now acidified, the precipitate of titanium(IV) hydroxide dissolves, and the excess titanium(III) is titrated with standard iron(III) chloride to a thiocyanate end point [4].

Method 2. Dissolved oxygen is titrated coulometrically with chromium(II) ion electrolytically generated in the solution being investigated. A biamperometric end-point detection is used. A pool of mercury acts as generator cathode. The generator anode consists of platinum foil in saturated potassium chloride solution. Potential of the generator cathode against SCE varies between 0.85 and 0.87 V. The error is less than 2% at concentrations as low as 0.03 ppm of oxygen [5].

Method 3. Trace concentrations of oxygen from 0.0008 to 0.6% in copper, molybdenum, tungsten, or lead are reduced with a continuous stream of pure hydrogen. The water so formed is transported by the gas, absorbed in dry methanol, and titrated with the Karl Fischer reagent [6].

Method 4. Oxygen dissolved in water is determined indirectly by a reaction involving the oxidation of manganese(II) hydroxide by the dissolved oxygen. The manganese(III) is complexed in acid solution with pyrophosphate following which the mixture is titrated visually with 0.05 N hydroquinone using diphenylamine as the indicator. In samples containing only microconcentrations of oxygen, the end point is determined potentiometrically or biamperometrically using platinum indicator electrodes versus SCE. The relative error is about 5% for the determination of 0.5 mg oxygen per liter with visual detection of the end point. Amounts of oxygen as low as 10 µg may be determined potentiometrically or biamperometrically. The method may also be adapted to the determination of dissolved oxygen in the presence of iron(II) or dissolved chlorine [7,8,9].

References for Elemental Oxygen

[1]. N. H. Furman, ed., *Standard Methods of Chemical Analysis*, 6th Ed., Vol. 1, D. Van Nostrand, Princeton, N. J., 1962, pp. 784-786.

[2]. L. W. Winkler, *Ber.*, *21*, 2843 (1888).

[3]. L. B. Pepkowitz and E. L. Shirley, *Anal. Chem.*, *25*, 1748 (1953).

[4]. E. S. Boĭchinova, S. M. Efros, and V. D. Nemirovskiĭ, *Trudy Leningrad. Tekhnol. Inst. im. Lensoveta, 1959, 58*, 31-35.

[5]. G. S. James and M. J. Stephen, *Analyst, 85*, 35-39 (1960).

[6]. W. Fisher, H. Bastius, and R. Mehlkorn, *Acta Chim. Acad. Sci. Hung.*, *34*, 167-178 (1962).

[7]. A. Berka and P. Hofmann, *Chem. Průmysl 13*, 287-290 (1963).

[8]. A. Berka, H. Glassl, and P. Hofmann, *Mikrochim. Acta 1967*(5), 828–833.

[9]. A. Berka, H. Glassl, and P. Hofmann, *Mikrochim. Acta 1968*(7), 997–1002.

Synoptic Survey of Water Determinations

The primary method for the determination of water is by redox titrimetry.

Classical Methods

Redox Titration. Small amounts of water in nonaqueous media are titrated with the Karl Fischer reagent [1,2].

Contemporary Methods

Redox Titration. Small amounts of water in nonaqueous media are titrated coulometrically with the Karl Fischer reagent. Iodine is continually regenerated in the depleted reagent [3].

Outline of Recommended Procedures for the Determination of Water

Redox Titration

Traces of water in organic solvents are titrated with coulometrically generated Karl Fischer reagent. For this purpose, iodine is electrolytically generated in a solution of the depleted reagent in ethylene glycol. The titration is carried out using a polarized platinum wire indicator electrode system and a platinum gauze generator electrode system. For titrations in methanol, a dead-stop technique is used for detection of the end point. The effect of side reactions which consume iodine is eliminated by a supplementary generating current adjusted to maintain the solution at the end point before the addition of the sample. The stoichiometry of the reaction shows that 1 mole of iodine is equivalent to 1 mole of water, or 10.71 coulombs of generating current per 1 mg of water. Samples containing as little as 5 γ of water per milliliter of solvent may be analyzed. The absolute standard deviation is approximately 2 γ of water [3].

References for Water

[1]. F. Fischer, *Angew. Chem.*, *48*, 394 (1935).

[2]. J. Mitchell and D. M. Smith, *Aquametry*, 2nd. Ed., Wiley-Interscience, New York-London, 1968.

[3]. A. S. Meyer and C. M. Boyd, *Anal. Chem.*, *31*, 215–219 (1959).

Synoptic Survey of Peroxide Determinations

Acceptable procedures for the determination of hydrogen peroxide and peroxycompounds include redox titrations.

Classical Methods

Redox Titrations. Hydrogen peroxide is determined iodometrically with standard thiosulfate [1].

Hydrogen peroxide, inorganic and organic peroxides, and, in general, peroxycompounds containing the -O-O- group can be titrated oxidimetrically with strong oxidizing agents such as permanganate and cerium(IV) salts [2,3].

Contemporary Methods

Hydrogen peroxide is titrated with standard hypobromite [4].

Hydrogen peroxide is determined by the addition of excess thiosulfate and the excess is back-titrated with coulometrically generated iodine [5].

Hydrogen peroxide and derivatives are titrated with ferrous solutions. The iron(III) formed is used to oxidize the leuco-base of methylene blue which in turn is titrated with standard vanadium(II) chloride [6].

Hydrogen peroxide is determined by coulometric cerimetry [7,8,9] or by amperometric titration with permanganate or dichromate [10].

Outline of Recommended Contemporary Methods for Hydrogen Peroxide and Peroxycompounds

Redox Titrations

Method 1. Hydrogen peroxide is determined in alkaline bromide media by titration with standard hypochlorite. The reactions involved are:

$$OCl^- + Br^- \rightarrow OBr^- + Cl^-$$

$$H_2O_2 + OBr^- \rightarrow Br^- + H_2O + O_2$$

The end point is detected photometrically by monitoring the absorbance of hyprobromite at 333 mμ during the analysis of samples containing 0.1-1.0 meq of hydrogen peroxide. For larger amounts of oxygen, Bordeaux red which is destructively oxidized at the end point by excess NaOBr is used as the indicator. The relative standard deviation for the titration is of the order of 0.2% [4].

Method 2. Hydrogen peroxide is determined by adding the sample to a solution of potassium iodide in the presence of an ammonium molybdate catalyst. An excess of standard thiosulfate is added, and the excess is determined by electrolytically generated iodine. The iodine is generated at 4.825 mA. The generating anode and cathode are made of platinum foils 1.5 cm^2 and 0.8 cm^2, respectively. The indicating

electrodes are made of two platinum foils 2 cm^2 in area with 135 mV
impressed between them. Best results are obtained at a pH 2.5-4.5.
The end point is found amperometrically. In the determination of 2.6
mg to 0.9 gm hydrogen peroxide, the average relative error ranges from
less than 0.1% to 4% [5].

Method 3. Hydrogen peroxide and other peroxy compounds such as
BaO_2 and $(NH_4)_2S_2O_8$ are determined by adding an excess of a ferrous
salt solution plus the leuco base of methylene blue. The resulting
ferric ions generated by the reaction of Fe(II) and the peroxide or
peroxy compound oxidize an equivalent amount of the leuco base to
methylene blue. The methylene blue so formed is then titrated with a
standard solution of vanadium(II) chloride to a colorless end point [6].

Method 4. Samples containing 50 μeq of hydrogen peroxide are
titrated with cerium(IV) which is electrogenerated at a constant current of 9.80 mA. The accuracy is 2-5% [7,8,9].

Method 5. Hydrogen peroxide is titrated amperometrically with
0.1 N potassium permanganate or 0.1 N potassium dichromate using a
platinum wire electrode versus SCE. The current direction remains
constant during the titration. A linear titration curve is observed
in the region of the equivalence point [10].

References for Hydrogen Peroxide and Peroxycompounds

[1]. I. M. Kolthoff, *Z. Anal. Chem.*, *60*, 400 (1921).

[2]. E. C. Hurdis and H. Romeyn, *Anal. Chem.*, *26*, 320 (1954).

[3]. E. E. Huckaba and F. G. Keyes, *J. Am. Chem. Soc.*, *70*, 1640 (1948).

[4]. W. H. McCurdy, Jr. and H. F. Bell, *Talanta*, *13*(7), 925-928 (1966).

[5]. G. D. Christian, *Anal. Chem.*, *37*(11), 1418-1420 (1965).

[6]. L. A. Ketova and S. I. Gusev, *Sb. Nauchn. Tr. Permsk. Gos. Med.
      Inst. 1962*, *41*, 145-147.

[7]. T. Takahashi and H. Sakurai, *Rep. Inst. Ind. Sci. Univ. Tokyo*,
      13(1), 1-37 (1963).

[8]. H. Sakurai, *Kagyo Kagaku Zasshi*. *64*, 2119 (1961).

[9]. T. Takahashi and H. Sakurai, *Talanta*, *9*, 189 (1962).

[10]. E. Michalski and N. Pawluk, *Chem. Anal. (Warsaw)*, *11*(5), 917-
       922 (1966).

SULFUR

In this section will be discussed methods for the determination
of elemental sulfur, hydrogen sulfide and sulfide ion, sulfite, sulfur
trioxide, sulfuric acid and sulfates, thiosulfate, dithionite, tetrathionate, polythionate, thiocyanate, sulfur monochloride, sulfur dichloride, and carbon disulfide.

Synoptic Survey of Elemental Sulfur Determinations

Methods for the determination of elemental sulfur involve precipitation titration.

Classical Methods

None of importance.

Contemporary Methods

Precipitation Titration. Sulfur reacts with potassium cyanide
to form potassium thiocyanate. Any excess cyanide is masked with
formaldehyde, and the thiocyanate is then titrated amperometrically
with standard silver nitrate [1].

Outline of Recommended Contemporary Methods for Elemental Sulfur

Precipitation Titration

Samples containing about 0.3 gm of elemental sulfur are treated
in 50% isopropyl alcohol solution with about 2 gm of potassium cyanide.
The solution is heated to form the thiocyanate. After cooling, the
excess cyanide is masked with formaldehyde, dilute nitric acid and
0.5% gelatin solution are added, and the thiocyanate ion is titrated
amperometrically with standard silver nitrate using a rotating platinum
indicator electrode without polarizing voltage [1].

Reference for Elemental Sulfur

[1]. S. Ikeda and S. Musha, *Bunseki Kagaku.* 15(8), 871 (1966).

Synoptic Survey of Hydrogen Sulfide and Sulfide Ion Determination

Methods for the determination of hydrogen sulfide and the sulfide ion include redox, complexometric, and precipitation titrations.

Classical Methods

None of importance.

Contemporary Methods

Redox Titrations. Traces of hydrogen sulfide are titrated coulometrically with iodine. The end point is determined amperometrically
[1].
Sulfide is determined by oxidation with excess calcium hypochlorite; the excess hypochlorite is removed by the addition of standard
arsenious acid, again in excess, and finally the amount of excess arsenious acid is titrated with standard 0.05 N Ca(OCl)$_2$ [2].

Complexometric Titration. Inorganic sulfides and mercaptans are
determined by high frequency conductometric titration in nonaqueous
solvents with mercuric salts [3].

Precipitation Titrations. Mixtures of hydrogen sulfide and low
molecular weight mercaptans are titrated potentiometrically with 0.1
N silver nitrate in an aqueous electrolyte solution [4].
Sulfide is determined in the presence of cyanide by an argento-
metric titration [5].
Sulfide ion is titrated potentiometrically with lead nitrate [6].

Outline of Recommended Contemporary Methods for Sulfide Ion

Redox Titrations

Method 1. Traces of hydrogen sulfide are titrated with coulometrically generated iodine. The method is easily adaptable to automatic titration methods. Samples containing 1-50 µg of sulfur in 10
milliliters may be determined with an accuracy of ± 1.5%. The method
is recommended for the determination of sulfur in hydrocarbons after
hydrogenation [1].

Method 2. Sulfide ion is determined by oxidation to sulfate
with excess calcium hypochlorite in about 5 N sodium hydroxide solutions. After standing 5-10 minutes to complete the oxidation, 0.05 N
arsenious acid is added followed by 10 N hydrochloric acid to a bromothymol blue end point. The mixture is then buffered with borax, a few
milliliters of 20% aqueous potassium bromide is added, and the mixture
is titrated with 0.05 N calcium hypochlorite. The difference between
the total hypochlorite and the amount of arsenious acid determines
the amount of sulfide [2].

Complexometric Titration

Samples containing 2-20 µeq of sulfide are determined by high
frequency conductometric titration in nonaqueous solvents such as
ethyl alcohol, acetone etc. with a standard solution of mercury(II)
chloride, acetate, nitrate, or thiocyanate. Mercaptans including
mixtures of mercaptans and sulfides in the presence of chlorides can
be analyzed by this method. The method is quite accurate and more
convenient and rapid than usual potentiometric methods. Since the
ionic strength of the solution affects the results, the determination
of traces of sulfide in aqueous solutions is not recommended [3].

Precipitation Titrations

Method 1. Mixtures of hydrogen sulfide and low molecular weight mercaptans are titrated potentiometrically with 0.1 N silver nitrate in an aqueous electrolyte which is 1 M in sodium hydroxide and 0.05 M in ammonium hydroxide. For higher molecular weight mercaptans in the presence of hydrogen sulfide, and in a solvent composed of sodium acetate dissolved in ethanol, coprecipitation affects the accuracy of the determination. However, the individual components of such a mixture may be determined since the mercaptan may be titrated in a second sample from which the hydrogen sulfide has been removed by precipitation as cadmium sulfide. The titration is carried out using a silver rod indicating electrode and a glass reference electrode or a 0.1 N sodium acetate reference electrode connected to the titration cell by a salt bridge and a galvanometer to read the null point. With a glass electrode, the observed potentials are more positive by about 150 mV. There are no interferences from iodide, bromide, or cyanide ions [4].

Method 2. In a mixture of sulfide and cyanide ions, sulfide ion is determined by argentometric titration to a potentiometric end point in a strongly alkaline ammoniacal solution. A rotating silver sulfide-silver indicator electrode versus SCE is used. The potential equilibrium is attained rapidly, and the potential break in the vicinity of the equivalence point is very large. Samples containing 0.2-4.0 mg sulfide in the concentration range of 5×10^{-4} to 1×10^{-2} M can be determined with a relative error of 2-4 parts per thousand. The simultaneous determination of sulfide and cyanide is possible since the titration curve shows two distinct breaks. The relative standard deviation is about 0.4%. The silver sulfide-silver electrode shows reversible behavior in the titration medium and, thus, acts as an efficient and adequate indicator electrode. Reducing species such as sulfite, sulfur, and thiocyanate do not interfere [5].

Method 3. Samples containing 1×10^{-3} to 1×10^{-5} mole of sulfide in 2 N sodium hydroxide solution are titrated with 0.05-0.005 M lead nitrate solutions. A silver wire indicator electrode rotating at 1000 rpm versus SCE reference electrode is used to determine potentiometrically the end point. Regular cleaning of the electrode surface is necessary to achieve rapid establishment of potential equilibrium. Sulfate, carbonate, thiocyanate, cyanide, chloride, thiosulfate, and ferrocyanide do not interfere [6].

References for Hydrogen Sulfide and Sulfide Ion

[1]. M. Pribyl and Z. Slovak, *Mikrochim. Ichoanal. Acta 1963*, (5-6), 119-125.

[2]. E. J. de la Pena and S. F. de Fazzini, *Rev. Fac. Ing. Quim. Univ. Nac. Litoral, Santa Fe, Argent. 33-34*, 197-199 (1964-1965).

[3]. L. Serrano Berges and T. Fernandez Perez, *An. Real Soc. Espan. Fis. Quim., Ser. B, 62*(7-8), 793-806 (1966).

[4]. M. W. Tamele, L. B. Ryland, and R. N. McCoy, *Anal. Chem., 32,* 1007-1011 (1960).

[5]. C. H. Liv and S. Shev, *Anal. Chem., 36,* 1652-1654 (1964).

[6]. V. A. Kremer, E. I. Vail, and A. Ya. Miroshnik, *Byull. Tekh. Ekon. Inform. Gos. Nauch.-Issled. Inst. Nauch. Tekh. Inform. 19*(9), 12-13 (1966).

Synoptic Survey of Sulfite Determinations

Methods for the determination of sulfite include acid-base, redox, and complexometric titrations.

Classical Methods

Acid-Base Titration. Depending on the nature of the sample, sulfite is titrated with standard base or standard acid using the appropriate indicator [1].

Redox Titration. Sulfite may be determined by titration with a wide range of oxidizing agents such as iodine [2], ammonium vanadate [3], and sodium chlorite [4].

Contemporary Methods

Redox Titrations. Sulfite ion is titrated with standard potassium ferricyanide in alkaline solution in the presence of osmium tetroxide as a catalyst [5].

Sulfite and chlorate ions in a mixture may be determined by titration of the sulfite with standard hydrogen peroxide followed by reduction of the chlorate to chloride with excess ferrous sulfate and back-titration with standard permanganate [6].

Sulfite is titrated with electrolytically generated bromine [7].

Complexometric Titration. In a mixture containing sulfide and sulfite, the sum of the two components is determined by oxidation with bromine to sulfate, precipitation with excess standard barium ion solution, and back-titration with standard EDTA. In a second aliquot, the sulfite is precipitated by the addition of alcohol, removed by filtration, and titrated as before after oxidation with bromine. Sulfide is found by difference [8].

Outline of Recommended Contemporary Methods for Sulfite

Redox Titrations

Method 1. Sulfite ion is titrated with standard potassium ferricyanide in alkaline solution in the presence of osmium tetroxide as a catalyst; osmiates may also be used. This method has at least one advantage over the usual iodometric method in that there is no loss of sulfur dioxide from the basic titration medium. Furthermore, oxidation by atmospheric oxygen is negligible. The error is of the order of 0.2%. This titration with standard ferricyanide may also be used for the titration of sulfide, thiosulfate, and tetrathionate [5].

Method 2. In a mixture containing both sulfite ion and chlorate ion, the components may be determined by first oxidizing the sulfite ion with excess standard hydrogen peroxide. The excess is back-titrated with standard potassium permanganate. In the same solution, the chlorate ion is next reduced to chloride ion with excess standard ferrous sulfate, and the excess is determined by titration with standard potassium permanganate. The results of this procedure agree favorably with those obtained by an iodometric titration [6].

Method 3. Samples containing from about five to several hundred parts per million sulfite ion are titrated with electrolytically generated bromine from potassium bromide. The titration is carried out in a coulometric cell with two platinum electrodes in the detection circuit, and two platinum electrodes connected by a salt bridge in the generation circuit. The sensitivity is better than 2 ppm [7].

Complexometric Titration

Sulfite and sulfide are determined in a mixture by a double titration. One aliquot of the sample is treated with bromine in order to oxidize both constituents to sulfate. The sulfate is then precipitated by the addition of excess barium ion solution, and the excess is back-titrated with standard EDTA. In a second aliquot, the sulfite is precipitated by the addition of ethyl alcohol, removed by filtration, dissolved in water, and oxidized to sulfate with bromine. The sulfate thus produced from the sulfite but not the sulfide is determined as before. Sulfide is found by difference. The results are satisfactory. The method is simple and rapid and is to be recommended for industrial analytical procedures [8].

References for Sulfite

[1]. N. H. Furman, ed., *Standard Methods of Chemical Analysis*, 6th Ed., Vol. 1, D. Van Nostrand, Princeton, N. J., 1962, pp. 1017-1018.

[2]. N. H. Furman, ed., *Standard Methods of Chemical Analysis*, 6th Ed., Vol. 1, D. Van Nostrand, Princeton, N. J., 1962, pp. 1016-1017.

[3]. R. Lang and H. Kurtenacker, *Z. Anal. Chem.*, *123*, 169 (1942).

[4]. D. T. Jackson and J. L. Parsons, *Ind. Eng. Chem. Anal. Ed.*, *9*, 14 (1937).

[5]. F. Solymosi and A. Varga, *Acta Chim. Acad. Sci. Hung.*, *20*, 295–
      306 (1959).

[6]. I. E. Flis and T. A. Tumanova, *Zavodsk. Lab.*, *26*, 943–945
      (1960).

[7]. P. Gh. Zugravescu and M. A. Zugravescu, *Rev. Chim. (Bucharest)*,
      18(1), 51–52 (1967).

[8]. A. De Sousa, *Inform. Quim. Anal. (Madrid)*, *17*(2), 51–52 (1963).

Synoptic Survey of Sulfur Trioxide Determinations

Methods for the determination of sulfur trioxide involve acid-
base and precipitation titrations.

Classical Methods

None of importance.

Contemporary Methods

Acid–Base Titration. The sulfur trioxide content of oleum is
determined by a potentiometric titration with aqueous sulfuric acid
[1].

Precipitation Titration. Small amounts of sulfur trioxide in
the presence of large amounts of sulfur dioxide are determined by
absorption in aqueous isopropyl alcohol and titration with standard
barium perchlorate using thoron, sodium 1-(o-arsenophenylazo)-2-naph-
thol 3,6-disulfonate as the indicator.

Outline of Recommended Contemporary Methods for Sulfur Trioxide

Acid–Base Titration

Sulfur trioxide contained in oleum is determined by a potentiometric titration with 86.35% aqueous solution of sulfuric acid. The
reaction involved is:

$$H_3SO_4^+ \; + \; HSO_4^- \; \rightarrow \; 2 \; H_2SO_4$$

The end point is determined using an antimony indicator electrode
versus a glass electrode or an antimony concentration reference electrode. The deflection in the titration curves at the end point is
quite large, amounting to about 200 mV. The standard deviation is
less than 0.1%.

Precipitation Titration

Samples of gases containing 50–500 µg of sulfur trioxide in the
presence of as much as 2000–3000 ppm of sulfur dioxide are determined

by a quantitative solution of the sulfur trioxide in a mixture of 4:1 isopropyl alcohol : water. Aliquots of this solution are then titrated with 0.0025 M barium perchlorate using thoron as the metallochromic indicator. The color change at the end point is from yellow to pink and is best viewed in subdued light. The standard deviation is 2 µg of sulfur trioxide. The range of titration is extended by determining the end point photometrically. In this case, samples containing 8 µg of sulfur trioxide are dissolved and titrated with 0.0005 M barium perchlorate with a standard deviation of 0.6 µg. Sulfur dioxide in concentrations up to 60 µg does not interfere. The method is more precise and sensitive than the classical turbidimetric method, and is recommended for the analysis of flue gasses [2].

References for Sulfur Trioxide Determinations

[1]. L. Giuffre, E. Losio, and A. Castoldi, *Chim. Ind. 48*(9), 958-961 (1966).

[2]. R. S. Fielder and C. H. Morgan, *Anal. Chim. Acta, 23,* 538-540 (1960).

Synoptic Survey of Sulfuric Acid Determinations

Methods for the determination of sulfuric acid involve acid-base titration.

Classical Methods

Acid-Base Titration. Sulfuric acid is titrated with standard base [1].

Contemporary Methods

Acid-Base Titrations. Sulfuric acid in the presence of various other mineral acids is titrated potentiometrically with standard sodium hydroxide or standard piperidine [2].
Sulfuric acid in mixtures with various sulfonic acids is titrated differentially in acetone with standard quaternary ammonium hydroxides [3,4].

Outline of Recommended Contemporary Methods for Sulfuric Acid

Acid-Base Titrations

Method 1. Sulfuric acid in the presence of nitric, hydrochloric, perchloric, phosphoric, acetic, salicylic, or p-toluenesulfonic acid is titrated potentiometrically in ethylene glycol with standard sodium hydroxide or piperidine dissolved in isopropyl alcohol. The titration curve shows two distinct breaks so that the sulfuric acid may be titrated either as a mono- or as a dibasic acid. The titration is carried out using a glass electrode versus SCE. The following tertiary mixtures containing sulfuric acid can also be titrated: acetic, p-toluenesulfonic; phosphoric, salicylic; phosphoric, acetic; nitric, phosphoric; and perchloric, phosphoric acids. However, in some cases, the second and/or third inflection in the titration curve is poor and accurate analysis is not possible. The method can still be used when rapid analysis and not a high degree of accuracy is of primary importance. Sulfuric acid may also be titrated conductometrically with piperidine in 2:1 ethylene glycol : acetone mixtures. Two sharp breaks in the titration curve are observed. Pure glycol may also be used as a solvent but the titration curves in a 1:1 mixture of ethylene glycol : acetone are unsatisfactory [2].

Method 2. The constituents of mixtures of sulfuric acid with various sulfonic acids in acetone may be titrated differentially with standard tetraethylammonium hydroxide in a solvent consisting of two parts of benzene and one part of ethyl alcohol. A mixed indicator made of methyl red and thymolphthalein is used. The titration may also be carried out in acetone solution using tetra-n-butylammonium hydroxide and a mixed indicator composed of neutral red and thymolphthalein. There is close agreement between the results of this procedure and the results obtained potentiometrically or gravimetrically [3,4].

References for Sulfuric Acid

[1]. B. J. Heinrich, M. D. Grimes, and J. E. Puckett, *Sulfur*, in *Treatise on Analytical Chemistry* (I. M. Kolthoff and P. J. Elving, eds.), Vol. 7, Interscience, New York-London, 1961, pp. 100-101.

[2]. M. N. Das and D. Mukherjee, *Anal. Chem.*, *31*, 233-237 (1959).

[3]. E. A. Gribova, *Zavodsk. Lab.*, *27*, 154-157 (1961).

[4]. W. Carasik, M. Mausner, and G. Spiegelman, *Chem. Spec. Mfr. Asc. Proc. Ann. Meet.*, *53*, 97-99 (1966).

Synoptic Survey of Sulfate Determinations

Methods for the determination of sulfate include acid-base, redox, complexometric, and precipitation titrations.

Classical Methods

Acid-Base Titration. Sulfate is precipitated as benzidine sulfate followed by titration with standard base [1].

Precipitation Titrations. Sulfate is precipitated with a slight excess of standard barium ion followed by titration with standard potassium chromate [2].

Sulfate is precipitated by barium chromate. The equivalent
amount of chromate liberated is determined by iodometric titration [2].

Sulfate is titrated with standard barium chloride using rhodizon-
ic acid or tetrahydroquinone as the indicator [3].

Contemporary Methods

Redox Titration. In the presence of phosphate, an acidic sulfate
solution is equilibrated with solid barium iodate in an acetone-acetic
acid medium. The liberated iodate is determined iodometrically [4].

Complexometric Titration. Sulfate is precipitated as lead sulfate, dissolved in excess standard EDTA and the excess back-titrated
with standard zinc chloride [5].

Precipitation Titrations. Sulfate is precipitated with excess
standard barium acetate, and the excess is back-titrated conductometrically with standard perchloric acid in 10% acetic anhydride [6].

Sulfate is titrated with barium perchlorate using dimethylsulfonazo III, 4,5-dihydroxy-3,6-bis(p-methyl-o-sulfo-phenylazo)-naphtha-
lene-2,7-disulfonic acid, as the indicator [7,8].

Outline of Recommended Contemporary Methods for Sulfate

Redox Titration

The precipitation titration of sulfate as barium sulfate in the
presence of phosphate ion is difficult due to the formation of slightly
soluble barium phosphate. The selectivity in the precipitation may be
improved by increasing the acidity of the medium, and by decreasing
the barium ion concentration. One method of accomplishing this is to
use a slightly soluble barium salt such as barium iodate and equili-
brating it with an acidic solution containing the sulfate and phosphate
ions. The metathesis reaction liberates iodate ions which are then
determined iodometrically. The equilibration of the sulfate solution
with barium iodate is best performed in a mixture of 40% acetone and
0.1 M acetic acid in water. As little as 0.01 mmole of sulfate per
25 ml can be determined provided that the temperature is controlled at
$25 \pm 0.5°$. At this concentration, the relative standard deviation is
2%. At higher concentrations of sulfate ions, the precision is better
than 1%. Chloride, bromide, perchlorate, and nitrate in concentrations
of 0.3 M or less do not interfere. Phosphate ion concentrations must
be kept lower than 0.02 M otherwise results are high. Cations which
form insoluble iodides or which can oxidize iodide to iodine interfere
but can be eliminated prior to analysis by ion exchange procedures [4].

Complexometric Titration

Sulfate is precipitated from boiling aqueous ethyl alcohol solutions as lead sulfate. The precipitate is removed and dissolved in
an excess of standard EDTA. The excess is then back-titrated with
standard zinc chloride solution using Eriochrome Black T as the indicator. The maximum allowable nitric acid concentration is 6%.

Phosphate, molybdate, and selenate interfere; arsenic and antimony do
not. Precision and accuracy are both very good. Since the time needed
for this titration is considerably shorter than that required for gravimetric procedures, the method is well-suited for routine purposes [5].

Precipitation Titrations

Method 1. Sulfate is precipitated with excess standard barium
acetate solution in an acetic acid medium. The excess is back-titrated
potentiometrically or conductometrically with standard perchloric acid
solution in 10% by volume of acetic anhydride. The conductometric
procedure is to be preferred over the potentiometric method since, in
the former case, water does not have to be completely excluded from the
sample. The titration is carried out using the usual conductance
bridge and two platinum electrodes 2 x 2 cm fixed 1 cm apart. The
precision and accuracy of the potentiometric and conductometric procedures are comparable [6].

Method 2. Sulfate is precipitated with 0.01 M barium perchlorate
using dimethylsulfonazo III as the indicator. Recoveries using 1.74-
13.94 mg potassium sulfate are $99.8 \pm 0.08\%$. The color change of the
indicator at the end point is from red to green-blue. Copper(II),
nickel, cobalt(II), zinc, iron(II), yttrium(III), lanthanides, pallad-
ium(II), and ruthenium(IV) interfere. The end point may also be determined spectrophotometrically at 655 mμ at which point the molar absorp-
tivity of the barium-dimethylsulfonazo III complex is 13,900 $cm^2 mmole^{-1}$
(in water) or 28,800 $cm^2 mmole^{-1}$ (in 2:3 acetone : water) at pH 7.0
[7,8].

References for Sulfate

[1]. N. H. Furman, ed., *Standard Methods of Chemical Analysis*, 6th
     Ed., Vol. 1, D. Van Nostrand, Princeton, N. J., 1962, pp. 1013-
     1014.

[2]. N. H. Furman, ed., *Standard Methods of Chemical Analysis*, 6th
     Ed., Vol. 1, D. Van Nostrand, Princeton, N. J., 1962, pp. 1012-
     1013.

[3]. W. Wagner, C. J. Hull, and G. E. Markle, *Advanced Analytical
     Chemistry*, Reinhold, N. Y., 1956, p. 207.

[4]. B. Jaselskis and S. F. Vas, *Anal. Chem.*, *36*(10), 1965-1967
     (1964).

[5]. A. W. Ashbrook and G. M. Ritcey, *Analyst*, *86*, 740-744 (1961).

[6]. G. Goldstein, D. L. Manning, and H. E. Zittel, *Anal. Chem.*, *34*,
     1169-1170 (1962).

[7]. B. Budesinsky and D. Vrzalova, *Chemist-Analyst*, *55*(4), 110
     (1966).

[8]. B. Budesinsky and L. Krumlova, *Anal. Chim. Acta,* *39*(3), 375-381
(1967).

Synoptic Survey of Thiosulfate Determinations

Thiosulfate is usually measured by means of redox titration.

Classical Methods

Redox Titration. Thiosulfate is titrated with iodine in neutral
solution [1].

Contemporary Methods

Redox Titrations. Thiosulfate is titrated with standard potassium ferricyanide in acid solution in the presence of zinc sulfate, potassium iodide, and starch [2].
     In a mixture of thiosulfate, sulfide, and polysulfide, thio-
sulfate is determined iodometrically after removal of sulfide and polysulfide with ammoniacal zinc sulfate [3].
     In a mixture of thiosulfate and sulfite, the sum of the two ions
is determined by potentiometric titration with iodine. In a second
aliquot, sulfite is masked with formaldehyde and the thiosulfate is
titrated potentiometrically again with iodine [4].
     In a mixture of thiosulfate and sulfide, the sulfide is removed
as cadmium sulfide; the thiosulfate is oxidized with bromine to sulfate, precipitated with excess standard barium chloride; and the unreacted barium chloride is back-titrated with standard EDTA [5].
     Thiosulfate is titrated potentiometrically with standard lead(IV)
tetraacetate [6].

Outline of Recommended Contemporary Methods for Thiosulfate

Redox Titrations

Method 1. Samples containing up to 0.5 meq thiosulfate in 70
ml are titrated with 0.05 M potassium ferricyanide in 0.1 M sulfuric
acid and in the presence of zinc sulfate, potassium iodide, and starch.
The reaction is catalyzed by the presence of the zinc sulfate. At
the end point the color change is from an essentially colorless solution to a light blue. The titration is rapid, and both precision and
accuracy are very high. The optimum concentration of acid is 0.1-0.4
N and of zinc 0.007-0.17 N. If the concentration of zinc is lower,
the liberation of iodine is not instantaneous, and if the concentration
is greater the results are low. If the thiosulfate concentration is
more than 0.005 M in the solution being titrated, thiosulfate is often
decomposed [2].

Method 2. Mixtures containing sulfide, polysulfide, and thiosulfate are analyzed by removal of the sulfide and polysulfide by the
addition of an ammoniacal zinc sulfate solution. Polysulfide is then
determined by titration of the sulfate formed according to the reactions:

$$Na_2S \cdot S_x \ + \ Zn(NH_3)_6(OH)_2 \ \rightarrow \ Zn(NH_3)_6(S \cdot S_x) \ + \ 2 \ NaOH$$

$$Zn(NH_3)_6(S \cdot S_x) \ + \ 8 \ HCl \ \rightarrow \ ZnCl_2 \ + \ H_2S \ + \ xS \ + \ 6 \ NH_4Cl$$

$$S \ + \ 2 \ KOH \ + \ 3 \ H_2O_2 \ \rightarrow \ K_2SO_4 \ + \ 4 \ H_2O$$

Thiosulfate is then determined iodometrically in the filtrate after precipitation of zinc sulfide. The sulfide may be determined indirectly by an iodometric titration of the alkali polysulfide solution. Total sulfur may also be determined in a separate aliquot by oxidation with bromine followed by precipitation of the sulfate with barium ion. Excess barium ion is used and the unreacted amount is back-titrated with standard EDTA using the mixed indicator Eriochrome Black T and sodium rhodizonate [3].

Method 3. Samples containing both sulfite and thiosulfate are analyzed by a potentiometric titration with iodine. In one aliquot, the sum of both ions is determined by titration with standard iodine, the end point being determined potentiometrically. In a second aliquot, the sulfite is masked by the addition of formaldehyde, and the thiosulfate alone is titrated potentiometrically with iodine. Sulfite is found by difference. The mean error in the determination of the sum of the two components is of the order of 3% [4].

Method 4. Samples containing both sulfide and thiosulfate are readily analyzed by precipitation of the sulfide as cadmium sulfide. The thiosulfate present is then oxidized to sulfate by the addition of bromine. The sulfate so formed is precipitated by the addition of an excess of standard barium chloride solution, and the unreacted barium ion is titrated with standard EDTA. In another aliquot the total sulfide and thiosulfate may be determined by oxidation with bromine and titration of the sulfate as before. The method is rapid and results are very satisfactory [5].

Method 5. Samples containing 5-100 mg thiosulfate in an aqueous solution containing sodium acetate are titrated potentiometrically with 0.0005-0.5 M lead(IV) tetraacetate. The titration may be direct or indirect. The electrode pair consists of a platinum rod versus SCE. The thiosulfate is oxidized quantitatively to the tetrathionate. The potential change at the equivalence point is about 70 mV for 0.2 ml of 0.05 M lead tetraacetate. The relative error is about + 0.67%. The titration may also be carried out by the addition of an excess of lead tetraacetate to the sample to oxidize the thiosulfate to sulfate. The unreacted lead tetraacetate is then titrated potentiometrically with standard hydroquinone [6].

References for Thiosulfate

[1]. N. H. Furman, ed., *Standard Methods of Chemical Analysis*, 6th Ed., Vol. 1, D. Van Nostrand, Princeton, N. J., 1962, pp. 1016-1017.

[2].   V. P. R. Rao and B. V. S. Sarma, *Chemist-Analyst, 55*(4), 110,
       125-126 (1966).

[3].   L. Legradi, *Analyst, 86*, 854-856 (1961).

[4].   V. K. Dubovaya and A. D. Miller, *Nekotorye Vopr. Khim. Teknol. i
       Fix. Khim. Analiza Neogran. Sistem, Akad. Nauk. Uz. SSR, Otd.
       Khim. Nauk, 1963*, 184-188.

[5].   A. De Sousa, *Inform. Quim. Anal. (Madrid), 17*(4), 139-141 (1963).

[6].   J. Rencova and J. Zyka, *Chemist-Analyst 56*(1-2), 27-28 (1967).

Synoptic Survey of Dithionite Determinations

Dithionite (hydrosulfite, hyposulfite) is determined by redox
titration.

Classical Methods

Redox Titration. Sodium dithionite is titrated with standard
indigo solution [1], or with iron(III) ion [2].

Contemporary Methods

Redox Titration. Dithionite is titrated with standard potassium
ferricyanide using methylene blue as the indicator [3].

Outline of a Recommended Contemporary Method for Dithionite

Redox Titration

Samples of commerically available dithionite are titrated in
alkaline solution with standard potassium ferricyanide using methylene
blue as the indicator. The color change at the end point is from
colorless to pale blue. The end point may also be detected potentiometrically. There are no interferences from the decomposition products
of dithionite such as sulfate, sulfite, sulfide, trithionate, or thiosulfate. Furthermore additives such as sodium chloride, phosphates,
soda, EDTA, urea, methanol, ethanol, or acetone do not interfere. For
commercial samples of sodium dithionite containing 91% sodium dithionite, the standard deviation using the above method is 0.3% [3].

References for Dithionite

[1].   N. H. Furman, ed., *Standard Methods of Chemical Analysis,* 6th
       Ed. Vol. 1, D. Van Nostrand, Princeton, N. J., 1962, pp. 1019-
       1020.

[2]. F. L. Hahn, *Anal. Chim. Acta*, *3*, 62 (1949).

[3]. D. C. De Groot, *Z. Anal. Chem.*, *229*(5), 335–339 (1967).

Synoptic Survey of Tetrathionate Determinations

Methods for the determination of tetrathionate involve redox titration.

Classical Methods

Tetrathionates are titrated iodometrically in the presence of other polythionates after treatment of aliquots of an aqueous solution containing tri-, tetra-, penta-, and hexathionates with sodium sulfite, potassium cyanide, sodium sulfide, and dilute sodium hydroxide [1].

Contemporary Methods

Redox Titrations. Samples containing tetrathionate are treated with excess standard hypochlorite. The unreacted hypochlorite is then back-titrated with standard thiosulfate [2].

Tetrathionate is titrated potentiometrically with standard potassium ferricyanide in the presence of osmium tetroxide as a catalyst [3].

Tetrathionate is treated with excess sulfite forming thiosulfate. The excess sulfite is masked with formaldehyde and the thiosulfate titrated iodometrically [4].

Outline of Recommended Contemporary Methods for Tetrathionate

Redox Titrations

Method 1. Samples containing tetrathionate are analyzed by oxidation with a basic solution of hypochlorite ion. An excess of potassium iodide is then added and the mixture is acidified with sulfuric acid. The liberated iodine is titrated with standard sodium thiosulfate. Dithionates, trithionates, and pentathionates are not quantitatively oxidized. The precision and accuracy are good [2].

Method 2. Tetrathionate is titrated potentiometrically with standard potassium ferricyanide in an alkaline solution and in the presence of osmium tetroxide as the catalyst. This direct determination of tetrathionate is reported to have much wider applications than previously reported methods [3].

Method 3. Samples containing 5×10^{-6} to 1×10^{-3} mole of tetrathionate ion are treated with excess sulfite ion in slightly alkaline solutions. The excess sulfite ion is masked by the addition of formaldehyde, and the thiosulfate formed in the redox reaction is determined iodometrically. The end point may be detected visually or spectrophotometrically at 372 mμ or 440 mμ. Accuracy is good. The following ions do not interfere: aluminum, calcium, cadmium, magnes-

ium, zinc, silicate, nitrate, sulfate, phosphate, chloride, or fluoride. Interferences are noted for iron(II), iron(III), manganese(II),
copper(II), and sulfite ions [4].

References for Tetrathionate

[1]. M. Goehring, U. Feldmann, and W. Helbing, Z. Anal. Chem., 129,
       346 (1949).

[2]. N. Hofman-Bang and M. T. Christiansen, Acta Chem. Scand., 15,
       2061 (1961).

[3]. F. Solymosi and A. Varga, Acta Chim. Acad. Sci. Hung. 20, 295-
       306 (1959).

[4]. I. Iwasaki and S. Suzuki, Bull. Chem. Soc. Japan, 39(3), 576-
       580 (1966).

Synoptic Survey of Polythionate Determinations

Polythionates are determined by acid-base or redox titration.

Classical Methods

Acid-Base Titration. Samples containing polythionates are
treated at pH 7 with saturated mercuric chloride solution. Excess
potassium iodide is added, and the hydrogen ion formed in the reaction
of the polythionate with mercuric chloride is titrated with standard
sodium hydroxide [1].

Contemporary Methods

Redox Titration. Polythionates are oxidized by the addition of
excess standard chloramine-T and the excess is titrated iodometrically
with standard thiosulfate [2].

Outline of a Recommended Contemporary Method for Polythionate

Redox Titration

Polythionates are determined by a procedure based on the rupture
of sulfur-sulfur bonds in the polythionate and the oxidation of the
sulfur to sulfuric acid by the reaction with excess chloramine-T solution. The sample containing polythionate is treated with excess 0.1
N chloramine-T in dilute acetic acid. The unreacted chloramine-T is
back-titrated by adding potassium iodide to the solution and titrating the liberated iodine with standard sodium thiosulfate. Dithionates do not interfere since they are not oxidized by the chloramine-T.

References for Polythionate

[1]. R. R. Jay, *Anal. Chem.*, *25*, 288-290 (1953).

[2]. K. Sharada and A. R. V. Murthy, *Z. Anal. Chem.*, *177*, 403-407
      (1960).

Synoptic Survey of Thiocyanate Determinations

Methods for the determination of thiocyanate include redox, complexometric, and precipitation titrations.

Classical Methods

Redox Titrations. Thiocyanate samples are treated in phosphoric
acid solution with a slight excess of bromine water. Potassium iodide
is added, and the liberated iodine is titrated with standard thiosulfate [1].
Thiocyanate is oxidized with iodine at pH 9-10 in a borate-boric
acid buffer. The excess iodine is titrated with sodium arsenite [2].

Contemporary Methods

Redox Titration. Thiocyanate is titrated with standard ceric
sulfate using iodine monochloride and ferroin as the indicator [3],
or using iodine monochloride alone as a catalyst and internal indicator [4].

Complexometric Titration. Thiocyanate is precipitated with silver ion. The silver thiocyanate is oxidized with nitric acid to sulfate which is precipitated with excess standard barium chloride. The
unreacted barium ion is then titrated with standard EDTA [5].

Precipitation Titration. Microamounts of thiocyanate are determined by high frequency titration with standard mercuric nitrate [6].

Outline of Recommended Contemporary Methods for Thiocyanate

Redox Titration

Thiocyanate is titrated with standard ceric sulfate solutions.
The ceric sulfate solution may be dissolved in sulfuric acid so that
the titrant is sulfatoceric acid. Iodine monochloride is added as a
preoxidizer and catalyst. Ferroin is employed as the indicator, or
the end point may simply be determined visually in which case the
iodine monochloride functions as the internal indicator. The end
point may also be determined biamperometrically. Results are excellent [3,4].

Complexometric Titration

Thiocyanate may be determined in the presence of cyanide and halide ion by precipitation of the ions in neutral solution as the insoluble silver salts. These are removed by filtration and boiled with nitric acid. The silver halides are unaffected, cyanide is expelled as HCN, and the thiocyanate is oxidized to sulfate. The solution is then filtered, and the sulfate is precipitated by the addition of excess standard barium chloride solution. The unreacted barium ion is then back-titrated with standard EDTA. Even very small amounts of thiocyanate can be determined in the presence of large amounts of cyanide and halide ions. Results are very satisfactory. The method is simple, rapid, and far more reliable than the usual argentimetric method which is complicated by the coprecipitation of cyanide and the halides [5].

Precipitation Titration

Very small amounts of thiocyanate in highly dilute solutions are easily and rapidly determined by high frequency conductometric titration with standard silver or mercuric nitrate. Standard commercial apparatus is used. The cell size is 100 ml. In the dilution ranges of 1/16,000 M to 1/32,000 M the error becomes about 2-4% respectively [6].

References for Thiocyanate

[1]. E. Schulek, *Z. Anal. Chem.*, *62*, 227 (1923).

[2]. B. R. Sant, *Ber.*, *88*, 581 (1955).

[3]. G. S. Deshmukh and A. L. J. Rao, *Z. Anal. Chem.*, *185*, 124, 376-377 (1962).

[4]. S. Singh and J. R. Siefker, *Anal. Chim. Acta*, *27*, 489-492 (1962).

[5]. A. De Sousa, *Inform. Quim. Anal. (Madrid)*, *14*(5), 130-131 (1963).

[6]. A. K. Mukherji and B. R. Sant, *Anal. Chim. Acta*, *20*, 295-262 (1959).

Synoptic Survey of Sulfur Monochloride Determinations

Sulfur monochloride is determined by redox titration.

Classical Methods

None of importance.

Contemporary Methods

Redox Titration. Sulfur monochloride is oxidized to sulfuric acid with excess chloramine-T. The unreacted chloramine-T is determined iodometrically [1].

Outline of Recommended Contemporary Methods for Sulfur Monochloride

Redox Titration

Samples of sulfur monochloride in dioxane are treated with an excess of acidified standard chloramine-T solution. Sulfur is oxidized to sulfuric acid. The excess chloramine-T is determined by the addition of an excess of 10% potassium iodide solution and titration of the liberated iodine with standard sodium thiosulfate. Precision and accuracy are very good [1].

Reference for Sulfur Monochloride Determinations

[1]. D. K. Padma and A. R. V. Murthy, *Talanta*, *12*(3), 295-297 (1965).

Synoptic Survey of Sulfur Dichloride

Sulfur dichloride is determined by precipitation titrimetry.

Classical Methods

None of Importance.

Contemporary Methods

Precipitation Titration. Sulfur dichloride is hydrolyzed and the liberated hydrochloric acid is determined argentimetrically [1].

Outline of Recommended Contemporary Methods for Sulfur Dichloride

Precipitation Titration

Samples of sulfur dichloride are hydrolyzed and the hydrochloric acid formed in the solution is titrated with standard silver nitrate. If present, sulfur dioxide and hydrogen sulfide interferences may be removed by oxidation with nitric oxide. The accuracy is reported to be within the range of 0.2-0.3% [1].

Reference for Sulfur Dichloride

[1]. S. A. Kiss, *Magyar Kém. Lapja*, *15*, 565-566 (1960).

Synoptic Survey of Carbon Disulfide Determinations

Methods for the determination of carbon disulfide include redox and complexometric titrations.

Classical Methods

Redox Titration. Carbon disulfide is dissolved in alcoholic potassium hydroxide and the resulting xanthate is titrated iodometrically [1].

Contemporary Methods

Redox Titration. Carbon disulfide is dissolved in alcoholic potassium hydroxide, and the xanthate so formed is titrated iodometrically at a specified time interval and at a pH of 5.7-5.8 with 0.1 N iodine using a starch indicator [2].

Complexometric Titration. Carbon disulfide is treated with diethylamine forming diethyldithiocarbamate which is titrated with standard mercuric chloride dissolved in pyridine [3].

Outline of Recommended Contemporary Methods for Carbon Disulfide

Redox Titration

The usual classical method for the determination of carbon disulfide involves formation of an alkali xanthate, $EtOCS_2K$, acidification with acetic acid and titration with iodine after the addition of bicarbonate. Satisfactory results are obtained only if an excess of bicarbonate is avoided, and if the titration is carried out in the cold with a freshly prepared iodine solution. Improvements in this procedure have recently been made and involve the use of 0.006-0.324 gm samples of carbon disulfide. The sample is dissolved in an excess of 10% alcoholic potassium hydroxide, and the excess is neutralized with acetic acid to a phenolphthalein end point. The pH is then adjusted to 5.7-5.8 with acetic acid or sodium acetate, and the mixture is titrated within a minute of acidification with 0.1 N iodine using a starch indicator. Both the time and the pH are important factors in obtaining accurate results. A wide pH range from 5.4-6.0 can be employed if an error of $\pm$ 1% can be tolerated [2].

Complexometric Titration

Samples of carbon disulfide are treated with diethylamine to form diethyldithiocarbamate. The carbamate is then titrated with 0.01 N mercuric chloride dissolved in pyridine using the copper-EDTA complex as the indicator. At the end point the brown solution is decolorized. All compounds which form insoluble products with mercury (II) ion and especially hydrogen sulfide and mercaptanes interfere. The method has definite advantages over the iodometric method since the procedure may be carried out in an aqueous medium, and the results are not dependent on pH or upon the time involved for the titration. The main disadvantage of this method is the narrow range of sample size which must be used. Samples must range from 0.5-15.0 mg of carbon disulfide in 30-35 ml of pyridine. This is due to the limited solubility of the product of the reaction of the carbon disulfide with diethylamine, the diethyldithiocarbamate, in aqueous pyridine. The accuracy is about the same as that of the iodometric method [3].

References for Carbon Disulfide

[1]. M. P. Matuszak, *Ind. Chem. Anal. Ed.*, *4*, 98 (1932).

[2]. M. Eusuf and M. H. Khundkar, *Anal. Chim. Acta*, *24*, 419-423 (1961).

[3]. V. Sedivec and J. Flek, *Collection Czechoslov. Chem. Communs. 24*, 3643-3648 (1959).

SELENIUM

Titrimetric methods for the determination of selenium include redox, complexometric, and precipitation titrimetry.

Synoptic Survey

Classical Methods

Redox Titrations. Selenium(IV) is oxidized to selenium(VI) with excess standard potassium permanganate. The excess is then back-titrated with standard ferrous ammonium sulfate [1,2].
Selenium(IV) is reduced to Se by excess potassium iodide in acid solution. The liberated iodine is titrated with standard thiosulfate [3].
Selenium(VI) is reduced with HCl to selenium(IV). The liberated chlorine is used to oxidize potassium iodide, and the liberated iodine is titrated with standard thiosulfate or arsenite [4].

Contemporary Methods

Redox Titrations. Selenium(IV) is titrated potentiometrically either with sodium hypochlorite or sodium hypobromite using osmium tetroxide as a catalyst. Visually the end point may be detected with diphenylamine sulfonic acid as the indicator [5].
Selenium(IV) is titrated coulometrically with electrogenerated iodide ion [6].
Microamounts of selenium(IV) are titrated coulometrically with electrolytically generated hypobromite ion [7].
Selenium(IV) is titrated with electrogenerated titanium(III) [8].

Complexometric Titration. Selenium metal is converted to $SeCN^{-1}$ by treatment with KCN. The excess cyanide is determined complexometrically [9].

Precipitation Titration. Selenium(IV) is reduced with hydrazine sulfate in the presence of excess standard silver nitrate forming silver selenide. The excess silver nitrate is titrated potentiometrically with standard potassium iodide [10].

Outline of Recommended Contemporary Methods for Selenium

Redox Titrations

Method 1. Samples containing 30-40 mg selenite are titrated potentiometrically (in a basic solution at 60-65°) with 0.1 N sodium hypobromite or 0.1 N sodium hypochlorite in the presence of osmium tetroxide as a catalyst. At the end point, the concentration of sodium hydroxide should not be lower than 1.5 M. The end point may also be established visually using diphenylamine sulfonic acid as the indicator. Using this indicator, the color change at the end point is from red to green. Accuracy is 0.2-0.4%. Ions that are oxidized by hypohalites in basic solutions interfere in the determination. Ethylenediamine, tellurates, and other substances which partially or completely inhibit the catalyst must, of course, be absent [5].

Method 2. Selenium(IV) is titrated coulometrically with electrogenerated iodide ion at constant current even in the presence of a fiftyfold excess of tellurium(IV). The end point is readily found amperometrically with two platinum polarized electrodes, $\Delta E = 0.1$ V, using a dead-stop technique and 1.43 mA. The sensitivity is 1.2×10^{-7} M or 0.4 γ selenium(IV) per milliliter. The relative error for samples containing 56-186 γ selenium(IV) is 1-4% with a relative mean deviation also of 1-4% [6].

Method 3. Microamounts of selenium(IV) are titrated coulometrically at constant current by electrolytically generated hypobromite ion. The end point is found potentiometrically. The sensitivity of the procedure is 2.8×10^{-5} M or 2.2 γ selenium(IV) per milliliter. Precision and accuracy are good [7].

Method 4. Samples containing 74-279 µg selenium(IV) are titrated coulometrically with electrogenerated titanium(III). The maximum current efficiency is obtained on platinum, gold, or tungsten electrodes in $\geq$ 12 N sulfuric acid solutions at 70° using $\geq$ 0.1 M titanium(IV) chloride over a wide current density range of 0.5-10 mA/cm^2. The end point may be determined potentiometrically or amperometrically using two polarized electrodes. If a potentiometrically determined end point is used, mixtures of selenium(IV) and tellurium(IV) can be analyzed successfully. The sensitivity for the selenium(IV) and tellurium(IV) determination is about 1×10^{-5} M [8].

Complexometric Titration

Selenium is converted to the selenocyanate ion ($SeCN^{-1}$) by treatment of the sample of metallic selenium with excess potassium cyanide. Excess standard nickel solution is then added followed by back-titration with EDTA using murexide as the indicator. The error is $\pm$ 1% [9].

Precipitation Titration

Samples containing 5-150 µg selenite are reduced with 2.5% hydrazine sulfate in the presence of an excess of standard silver nitrate solution. Silver selenide is precipitated. The excess silver nitrate is then titrated potentiometrically with standard potassium iodide. An accuracy of 1% is obtained even for samples containing very small amounts of selenium. This method may be used for the determination of selenium in the presence of tellurium provided that the tellurium to selenium ratio does not exceed 2 to 1 [10].

References

[1]. W. T. Schrenk and B. L. Browning, *J. Am. Chem. Soc.*, *48*, 2550 (1926).

[2]. S. Barabas and W. C. Cooper, *Anal. Chem.*, *28*, 129 (1956).

[3]. G. G. Marvin and W. C. Schumb, *Ind. Eng. Chem. Anal. Ed.*, *8*, 109 (1936).

[4]. C. S. Soth and J. E. Ricci, *Ind. Eng. Chem. Anal. Ed.*, *12*, 328 (1940).

[5]. F. Solymosi, *Chemist-Analyst*, *52*, 42-43 (1963).

[6]. L. B. Agasyan, P. K. Agasyan, and E. R. Nikolaeva, *Vestn. Mosk. Univ.*, *Ser. II*, *21*(5), 93-95 (1966).

[7]. L. B. Agasyan, E. R. Nikolaeva, and P. K. Agasyan, *Vestn. Mosk. Univ. Ser. II*, *21*(5), 96-98 (1966).

[8]. L. B. Agasyan, E. R. Nikolaeva, and P. K. Agasyan, *Zh. Anal. Khim.* *22*(6), 904-908 (1967). Engl. transl., *J. Anal. Chem. USSR*, *22*, 762-765 (1967).

[9]. D. Gimesi, G. Rady, and L. Erdey, *Acta Chim. Acad. Sci. Hung.* *33*, 381-385 (1962).

[10]. H. Hahn and H. Bartels, *Mikrochim. Acta*, *1961*, 259-261.

TELLURIUM

Methods for the determination of tellurium involve primarily redox titration.

Synoptic Survey

Classical Methods

Redox Titrations. Tellurium(IV) is oxidized to tellurium(VI)

in hot sulfuric acid solutions with excess standard ceric sulfate
followed by back-titration of the unreacted ceric ion with standard
ferrous ammonium sulfate [1].

Tellurium(IV) is titrated with standard permanganate solution
[2].

Contemporary Methods

Redox Titrations. Tellurium(IV) is determined coulometrically
with electrogenerated hypobromite ions [3].

Tellurium(VI) is reduced to tellurium (IV), reoxidized with a
known volume of standard potassium dichromate, and the excess dichromate is back-titrated with standard ferrous ion [4].

Tellurium(IV) is determined by amperometric titration with sodium hexyldithiocarbamate or sodium cyclopentyldithiocarbamate [5].

Tellurium(IV) is reduced to the metal which is then separated
and oxidized with excess standard ferric chloride. The ferrous ions
formed are titrated with standard ceric sulfate [6].

Outline of Recommended Contemporary Methods for Tellurium

Redox Titrations

Method 1. Microamounts of tellurium(IV) are determined coulometrically at constant current with electrolytically generated hypobromite ions. The end point is detected potentiometrically. The
sensitivity is 3.8 γ tellurium(IV) per milliliter. The titration is
best carried out at 70° in a solution which is 1 M in potassium bromide, 0.1 M in sodium bicarbonate, and at a current density of 0.5-5.0
mA/cm^2. Platinum electrodes, which are treated with nitric acid prior
to each determination, are employed. Also before each determination,
the medium must be preoxidized to the value of the potential at the
end point. This potential is + 0.55 V versus SCE. For samples containing 96-512 γ tellurium(IV), the relative error is 0-4%, and the
relative mean deviation is 1-7% [3].

Method 2. Tellurium(VI) is determined indirectly by reduction
to tellurium(IV) with boiling hydrochloric acid solutions. A known
excess of standard potassium dichromate solution is then added and
the unreacted dichromate is back-titrated with standard Mohr's salt
solutions using N-phenylanthranilic acid as the indicator [4].

Method 3. Samples containing tellurium(IV) are determined by
amperometric titration in aqueous hydrochloric acid at pH 4.0-5.5
with 0.01 M sodium hexyldithiocarbamate or sodium cyclopentyldithiocarbamate. These dithiocarbamates form insoluble stable derivatives
with selenium(IV) and tellurium(IV) in acid medium. The smooth oxidation of these derivatives on a platinum anode forms the basis for
the amperometric determination. Titrations are carried out at + 0.6-
0.8 V versus SCE. In samples containing both selenium and tellurium,
the total selenium and tellurium may be titrated at a pH lower than
3.6, while the tellurium in a separate sample is titrated at pH 4-5.5.
Selenium values can, of course, be obtained by difference. The

relative percent error is 1.6 for tellurium and 1.4 for selenium [5].

   Method 4. Samples containing tellurium(IV) are reduced to the metal by boiling with alkaline D-glucose solutions or solutions of other reducing sugars. The metal is then separated and oxidized with an excess of standard ferric chloride solution. The ferrous ions so formed are titrated with standard ceric sulfate using N-phenylanthranilic acid as the indicator. Lithium, silver, thallium(I), calcium, strontium, lead(II), palladium(II), magnesium, mercury(II), zinc, bismuth, gold(III), thorium(IV), zirconium(IV), and selenium(VI) interfere [6].

References

[1]. N. H. Furman, ed., *Standard Methods of Chemical Analysis*, 6th Ed., Vol. 1, D. Van Nostrand, Princeton, N. J., 1962, pp. 933-934.

[2]. H. H. Willard and P. Young, *J. Am. Chem. Soc.*, *52*, 553 (1930).

[3]. L. B. Agasyan, E. R. Nikolaeva, and P. K. Agasyan, *Vestn. Mosk. Univ.* [II], *21*, 51, 96-98 (1966).

[4]. N. P. Tikhomirova, *Metody Anal. Khim. Reaktivor Prep.*, *12*, 90-91 (1966).

[5]. F. M. Tulyupa, V. S. Baralov, and Yu I. Usatenko, *Zh. Anal. Khim.* *22*(3) 399-405 (1967). Engl. transl. *J. Anal. Chem. USSR*, *22*(3) 374-352 (1967).

[6]. O. C. Saxena, *Microchim. Acta, 1967*(6), 1123-1125.

POLONIUM

   Polonium is best determined by a titrimetric procedure involving oxidation and reduction.

Synoptic Survey

Classical Methods

   None reported.

Contemporary Methods

   Redox Titration. Polonium is determined by a coulometric titration in 0.1 M perchloric acid [1].

Outline of Recommended Contemporary Methods for Polonium

Redox Titration

Microgram samples of ^{210}Po from irradiated bismuth are freed of radiolytic degradation products by electrodeposition on gold electrodes at -0.4 V versus Hg-Hg$_2$SO$_4$ (in 1 M H$_2$SO$_4$) electrode. This electrode has a potential of 0.418 $\pm$ 0.003 V versus SCE. The polonium is then redissolved in 0.1 M perchloric acid and determined coulometrically. Intense alpha-bombardment leading to the production of hydrogen peroxide limits the titration to less than mono-layer amounts of ^{210}Po and the recommended amount is $\leq$ 0.1 µg/cm^2. Reproducible results are obtained when excess polonium is deposited and then stripped from the Hg-Hg$_2$SO$_4$ electrode to give the required concentration. The precision is poor and averages about 5% for samples containing 0.35-0.40 gm of ^{210}Po due to interfering radiation-produced products [1].

Reference

[1]. R. C. Propst, *Anal. Chem.*, *40*(1), 244-246 (1968).

CHROMIUM

Methods for the determination of chromium include redox and complexometric titrations.

Synoptic Survey

Classical Methods

Redox Titrations. Chromium(III) is titrated with standard potassium permanganate with the end point being determined visually or potentiometrically [1].
Chromium(VI) is reduced to chromium(III) with excess ferrous ions. The unreacted ferrous salt is back-titrated with standard dichromate. A visual or potentiometrically determined end point may be observed [2].
Chromium(VI) is reduced with iodide ion and the liberated iodide is titrated with standard thiosulfate [3].

Contemporary Methods

Redox Titration. Chromium(VI) is titrated potentiometrically in phosphoric acid solution with standard Mohr's salt solution [4].

Complexometric Titrations. Chromium(III) samples are used to reduce permanganate to manganese dioxide which, in turn, is dissolved in and further reduced by ascorbic acid. The manganese(II) ions so formed are then titrated with standard EDTA [5].

Chromium(VI) is reduced to chromium(III) with sodium acid sulfite in the presence of excess standard EDTA. The unreacted EDTA is back-titrated with standard thorium nitrate using xylenol orange as the indicator [6].

Outline of Recommended Contemporary Methods for Chromium

Redox Titration

Samples containing chromium(VI) are titrated potentiometrically in 6.5 M orthophosphoric acid solution with a standard Mohr's salt solution. The potentials of the systems Cr(VI)/Cr(III) and Fe(III)/Fe(II) are respectively 1.00 and 0.28 V versus SCE when measurements are made in 6.5 M H_3PO_4. Samples containing both chromium(VI) and vanadium(V) may also be titrated potentiometrically. Two potential changes of 200 mV and 100 mV are observed in the titration curve. The first corresponds to complete reduction of Cr(VI)→Cr(III) and V(V)→V(IV), the second to further reduction of V(IV)→V(III). The relative error is 1-2% for alloys containing 20-70 percent of the test element [4].

Complexometric Titrations

Method 1. Samples of chromium(III) in a neutral or weakly acid solution are treated with potassium permanganate solution in the presence of an acetate buffer. The redox reaction forms equivalent amounts of chromate ion and insoluble manganese dioxide. The manganese dioxide is removed and dissolved in ascorbic acid which completes the reduction to manganese(II). The latter is then titrated with standard EDTA. Relatively large amounts of zinc, cadmium, or copper do not interfere. Precision and accuracy are very good [5].

Method 2. Samples containing chromium(VI) are placed in a 0.2 M triethanolamine buffer adjusted to pH 6.6 with sulfuric acid and are reduced to chromium(III) with sodium acid sulfite in the presence of a slight excess of standard EDTA. The stable and nonreactive chromium(III)-EDTA complex is formed via the labile chromium(V) and chromium(IV) EDTA complexes. The excess EDTA is back-titrated at pH 2.5-3.5 with standard thorium nitrate using xylenol orange as the indicator. The time required for a single determination is only 5-6 minutes. The relative standard deviation is 1.4% for samples containing 0.5 mg chromium and 0.8% for samples containing 5.0 mg chromium [6].

References

[1]. N. H. Furman, ed., *Standard Methods of Chemical Analysis*, 6th Ed. Vol. 1, D. Van Nostrand, Princeton, N. J., 1962, pp. 354-355.

[2]. N. H. Furman, ed., *Standard Methods of Chemical Analysis*, 6th

Ed. Vol. 1, D. Van Nostrand, Princeton, N. J., 1962, p. 354.

[3]. N. H. Furman, ed., *Standard Methods of Chemical Analysis*, 6th
      Ed. Vol. 1, D. Van Nostrand, Princeton, N. J., 1962, pp. 355-
      356.

[4]. L. I. Veselago, *Zh. Anal. Khim.*, *23*(3) 384-387 (1968). Eng.
      transl. *J. Anal. Chem. USSR 23*(3) 317-319 (1968).

[5]. L. Szekeres, E. Kardos, and G. L. Szekeres, *Microchem. J.*, *12*(2)
      147-150 (1967).

[6]. D. A. Aikens and C. N. Reilly, *Anal. Chem., 34*, 1707-1709 (1962).

MOLYBDENUM

Methods for the determination of molybdenum include redox, complexometric, and precipitation titrimetry.

Synoptic Survey

Classical Methods

Redox Titrations. Molybdenum(VI) is titrated potentiometrically
with titanous ion [1].

Molybdenum(VI) is reduced with mercury to molybdenum(V) which is
titrated with standard ceric sulfate to an α-phenanthroline end point
[2].

Molybdenum(VI) is reduced to molybdenum(III) in a Jones reductor,
led into an excess of ferric sulfate solution, and the ferrous ions
so formed are titrated with standard permanganate [3].

Contemporary Methods

Redox Titrations. Molybdenum(V) is titrated with standard ceric
sulfate in hydrochloric acid solution using methyl orange as the indicator [4].

Molybdenum(VI) is titrated potentiometrically with standard ti-
tanium(III) chloride [5].

Molybdenum(VI) is reduced to molybdenum(V) with excess standard
iron(II) in concentrated H_3PO_4. The unreacted iron(II) is titrated
potentiometrically with standard permanganate [6].

Molybdenum(VI) is reduced with cadmium amalgam and titrated with
standard permanganate [7].

Molybdenum(VI) is reduced to molybdenum(V) with silver, mercury,
tin(II), or amalgamated zinc and the molybdenum(V) so formed is titrated with cerium(IV), dichromate, or metavanadate solutions [8].

Molybdenum(VI) is titrated coulometrically with electrogenerated
iron(II) [9], or titanium(III) [10].

Complexometric Titrations. Molybdenum(VI) is reduced in the presence of excess 0.01 M EDTA with hydroxylamine hydrochloride. The excess EDTA is back-titrated with standard bismuth solutions using xylenol orange as the indicator [11].

Molybdenum(VI) is titrated amperometrically with 8-mercaptoquinoline (thioxine) [12].

Molybdenum(VI) is titrated in 50% ethanol solutions with standard EDTA. The diphenylcarbazone-methylene blue-vanadium(V)-EDTA system is used for end-point detection [13].

Molybdenum(VI) is determined by high-frequency titration with cupferron [14].

Precipitation Titration. Molybdenum is titrated potentiometrically with standard silver nitrate [15].

Outline of Recommended Contemporary Methods for Molybdenum

Redox Titrations

Method 1. Molybdenum(V) is titrated with 0.1 N ceric sulfate at room temperature in 1 N hydrochloric acid using 0.2% methyl orange as the indicator. The titration may also be carried out in 1 N sulfuric acid using either methyl orange or methyl red as the indicator. At the end point in the hydrochloric acid medium, the indicator is irreversibly oxidized to yellow whereas in the sulfuric acid medium, the red color of the indicator fades rapidly and changes to yellow in 15-20 seconds after the addition of the last portion of titrant. The acid concentration is not critical. The accuracy is within 3 parts per thousand [4].

Method 2. Molybdenum(VI) is titrated potentiometrically with standard titanium(III) chloride in the presence of citric acid. For samples in the semimicro range, the standard deviation is $\pm$ 0.06 ml of 0.01 N titanium(III) chloride [5].

Method 3. Molybdenum(VI) is reduced to molybdenum(V) with an excess of standard ferrous solution in concentrated phosphoric acid (>11.5 M). The excess iron(II) is back-titrated potentiometrically with 0.2 N dichromate using platinum electrodes at a potential of 150-180 mV versus SCE. Two breaks are noted in the titration curve. At the first break, corresponding to the completion of the oxidation of the excess iron(II), an increase in potential of 30-90 mV per 0.04 ml of titrant is noted. At the second break, corresponding to the oxidation of the molybdenum(V), the increase in potential is 250-300 mV per 0.04 ml of titrant. Samples containing both molybdenum and vanadium may be analyzed by a slight variation in this method. Manganese (II), manganese(VI), copper(II), nickel(II), cobalt(II), zinc(II), iron(III), chromium(III), chromium(VI), cerium(III), cerium(IV), titanium(IV), thorium(IV), or tungsten(VI) do not interfere. Uranium and nitrate interfere. Precision and accuracy are excellent [6].

Method 4. Samples containing 0.025-0.4 gm molybdenum are reduced in 2-5 N sulfuric acid with 1% cadmium amalgam and titrated with standard potassium permanganate. Amounts of tungsten up to twofold may be tolerated provided that the tungsten is masked with fluoride ion [7].

Method 5. Samples containing molybdenum(VI) are reduced to molybdenum(III) by the Jones reductor. Since the molybdenum(III) is sensitive to atmospheric oxygen, the reduced solution is collected in a known excess of a standard oxidizing agent such as permanganate, cerium(IV), dichromate, or metavanadate which is then titrated with a reducing agent such as iron(II). Titrations are carried out in a solution acidified with orthophosphoric acid using ferroin, or N-phenylanthranilic acid as the indicator. The molybdenum found, as a percentage of that taken, is 98.98 + 0.01 with potassium dichromate as the oxidizing agent and 99.97 + 0.02 when the oxidizing agent is sodium metavanadate. The original reduction of the molybdenum(VI) to molybdenum(V) may be carried out using silver, mercury, or tin(II) as the reducing agent. Accuracy in all cases is excellent [8].

Method 6. Samples containing molybdenum(VI) are titrated coulometrically with electrogenerated iron(II). The optimum conditions for the generation of the iron(II) involves the use of platinum or tungsten electrodes, an iron(III) concentration of 0.1 M, a current density of 1 mA/cm^2, and a medium of 11 M orthophosphoric acid which is also 2 M in sulfuric acid. Amperometric detection of the end point is used. A single platinum-microelectrode (E = 0.8 V) or two polarized Pt/W electrodes (ΔE = 50-100 mV may be used). Sensitivity is 7 x 10^{-5} M Mo(VI) [9].

Method 7. Samples containing molybdenum may be titrated coulometrically with electrogenerated titanium(III) preceded by electrolysis of 0.7 M titanium(IV) chloride-8 M sulfuric acid solutions. Platinum electrodes with an applied potential of 250 mV are used. The recommended current density is less than 2 mA/cm^2. The titration must be carried out in a solution cooled to room temperature since the indicator current is temperature dependent. The average error is less than + 0.2% for samples containing 0.5-5 mg of molybdenum. Tin(IV), lead (II), cadmium(II), chromium(III), nickel(II), manganese(II), tungsten (VI), and magnesium(II) do not interfere. Interferences are to be noted for copper(II) and nitrate. Iron(III) when present in a weight ratio of less than 1:1 does not interfere [10].

Complexometric Titrations

Method 1. Molybdenum(VI) is reduced with hydroxylamine hydrochloride in the presence of EDTA. A molybdenum(V)-EDTA complex is formed over a pH range of 2-5 having a 1:1 Mo/EDTA ratio. Samples containing 5-25 mg amounts of molybdenum are reduced with 10% hydroxylamine hydrochloride in the presence of an aliquot of 0.01 M EDTA. The pH is then adjusted to pH 2 with ammonium sulfate, and, after boiling and cooling, the excess EDTA is back-titrated either with 0.01 M bismuth to a xylenol orange end point, or the titration is performed with 0.01 M copper solution at pH 4.5 using 1-(2 pyridylazo)-2-naphthol as the indicator [11].

Method 2. Molybdenum is titrated amperometrically with 8-mer-

captoquinoline(Thioxine) using a platinum indicator electrode and a
mercury(I) iodide reference electrode. The titration may be carried
out in different supporting electrolytes over a wide pH range. Titration curves are basically identical in 2 N, 1 N, and 0.1 N sulfuric,
nitric, or hydrochloric acids. Vanadium and chromium interfere. Twofold amounts of bismuth, tungsten, and tellurium can be tolerated.
The method is of value in the analysis of steel, titanium, aluminum,
and nickel alloys [12].

Method 3. Samples containing 5-500 mg molybdenum in 50% aqueous
ethanol are titrated at pH 5.2-5.6 (acetate buffer) with 0.05 M EDTA.
The titration is performed using methylene blue-diphenylcarbazone as
the mixed indicator, and the titration is continued until the solution
is colorless. One milliliter of vanadium(V)-EDTA complex is then
added and the titration is continued to a definite blue color at the
end point [10].

Method 4. Molybdenum(VI) is determined by high frequency titrations with cupferron. Molybdenum(VI) forms both 1:1 and 1:2 complexes
with cupferron on titration with 0.1 M cupferron. Two breaks are
obtained on the titration curve corresponding to their formation. If
the titration is carried out at a pH higher than 2.6, the inflection
at a molar ratio of 1:1 is still clearly defined, but the break at the
1:2 molar ratio tends to disappear. If the reaction is followed potentiometrically in a buffered solution at pH 4.6, only the 1:1 complex is formed. The experimental error does not exceed 0.5% and the
minimum concentration which can be determined is 5×10^{-5} M [14].

Precipitation Titration

Aqueous solutions of molybdenum are titrated at pH 7.2-7.6 with
standard silver nitrate. The precipitate has the composition Ag_2MoO_4.
Silver indicator electrode versus SCE reference electrode is used.
The accuracy is good even at very low molybdenum concentrations [15].

References

[1]. H. H. Willard and F. F. Fenwick, J. Am. Chem. Soc., 45, 928
(1923).

[2]. N. H. Furman and W. M. Murray, Jr., J. Am. Chem. Soc., 58, 1689
(1936).

[3]. N. H. Furman, ed., Standard Methods of Chemical Analysis, 6th
Ed., Vol. 1, D. Van Nostrand, Princeton, N. J., 1962, pp. 681-
682.

[4]. K. R. Rao, Rec. Trav. Chim., 84(1), 71-73 (1965).

[5]. H. Hahn and A. F. Moosmueller, Z. Anal. Chem., 221, 261 (1966).

[6]. U. Muralikrishna and G. G. Rao, Talanta, 15(1), 143-144 (1968).

[7].   E. F. Speranskaya and V. E. Mertsalova, *Khim i Khim. Tekhnol.*,
       Alma-Ata, Sb. 2, 89-91 (1964).

[8].   J. Becker and C. J. Coetzee, *Analyst, 92*, 166-169 (1967).

[9].   P. K. Agasyan, E. Kh. Tarenova, E. R. Nikolaeva, and R. M.
       Katina, *Zavod. Lab. 33*(5), 547-550 (1967).

[10]. Yu-Hui Yen and Yung-Hsiang Liu, *K'o Ksueh T'ung Pao 17*(6), 279-
       281 (1966).

[11]. H. Yaguchi and T. Kajiwara, *Bunseki Kagaku, 14*(9), 785-788
       (1965).

[12]. I. M. Pavlova and O. A. Songina, *Sb. Stat. Aspirantov Soiskatel.
       Min. Vyssh. Sredn. Spets, Obrazov Kaz. SSR, Khim, Khim. Technol.
       3-4*, 218-224 (1965).

[13]. I. Sajó, *Z. Anal. Chem., 199*(1), 16-19 (1963).

[14]. C. B. Riolo, T. F. Soldi, and G. Spini, *Anal. Chim. Acta, 41*,
       388-391 (1968).

[15]. G. C. Shivahare, *Fresenius' Z. Anal. Chem., 219*(2), 187-188
       (1966).

TUNGSTEN

Methods for the determination of tungsten include redox, complexometric, and precipitation titration.

Synoptic Survey

Classical Methods

None of the classical methods involving precipitation or redox
titrations can be recommended because of interference from various
other ions or because the methods give uncertain or varying results.

Contemporary Methods

Redox Titration. Tungsten(III) or tungsten(V) is titrated
with iodine monochloride solutions [1].

Complexometric Titration. Tungsten(VI) is precipitated as lead
tungstate with excess lead nitrate. The excess lead is back-titrated
with standard EDTA [2].

Precipitation Titration. Tungsten is titrated potentiometrically [3,4] or amperometrically [5] with standard lead nitrate.

Outline of Recommended Contemporary Methods for Tungsten

Redox Titration

A 10-20 ml aliquot of 0.1 N iodine monochloride in 3 N hydrochloric acid is diluted to 50 ml with water containing a few milliliters of concentrated phosphoric acid. The solution is degassed with
nitrogen and titrated potentiometrically with the unknown tungsten(V)
or tungsten(III) solution. The latter are prepared by reduction of
tungsten(VI) with zinc-amalgam or mercury respectively. The accuracy
is within $\pm$ 0.7-0.9%. This titration with iodine monochloride may
also be used for the determination of molybdenum(III) [1].

Complexometric Titration

Tungsten is determined indirectly by the addition of a known
excess of 0.1 M lead nitrate solution to a sample containing approximately 1 mg of tungstate per milliliter. The lead tungstate formed
is removed by filtration, and the excess lead is determined by titration with standard EDTA. As an alternate procedure, the unreacted
lead ions in the solution may be complexed with excess standard EDTA
and, in turn, the excess unreacted EDTA may be titrated with a standard zinc solution using Eriochrome Black T as the indicator [2].

Precipitation Titrations

Method 1. Samples containing tungsten(VI) are titrated potentiometrically in a neutral solution using tungsten-calomel or tungsten-
platinum electrodes. The titration is carried out at 50° with 0.05 M
lead nitrate. At the end point, which should be approached slowly,
the potential change is 100-150 mV per drop of the above titrant. The
end point may also be determined by means of an adsorption indicator
using 3-4 drops each of 0.2% amine black green B and 0.2% diamine
fast scarlet 6BS. The solution containing the tungstate is brought
to a boil, the above mixed indicator is added, and the solution is
titrated rapidly with 0.05 M lead nitrate until a color change occurs
and then dropwise until the solution turns blue and the precipitate
red with one drop of titrant. The error in either method is 0.1-0.2%
for samples containing 30 mg of tungsten [3,4].

Method 2. Tungsten may be determined in an alloy in the presence
of molybdenum by dissolving the sample in aqua regia, diluting with
water, followed by digestion for about an hour on a warm hot plate.
The tungstic acid precipitate is removed, dissolved in hot 10% sodium
hydroxide and an aliquot taken for analysis. The pH of the aliquot
containing 5-10 mg tungsten is adjusted to 4.5 using an acetate buffer,
ethanol is added, and the solution is titrated amperometrically at 80°
at -0.5 V with a lead nitrate solution using a rotating platinum electrode versus SCE. The error is $\pm$ 2-3% [5].

References

[1]. J. Sevik and J. Cihalik, *Colln. Czech. Chem. Commun. 31*(8), 3140-3153 (1966).

[2]. A. A. Bykovskaya, *Tr. Int. Met. A. A. Baikova, 1962, 10.*

[3]. H. Brubitsch, N. Ozbil, and K. Kluge, Z. *Anal. Chem., 166*, 114-120 (1959).

[4]. C. M. Gupta, *J. Proc. Inst. Chem. (India), 38*(5), 211-214 (1966).

[5]. V. I. Bogovina, V. P. Novak, and V. F. Mal'tev, *Zh. Analit. Khim., 20*(9), 951-954 (1965). Eng. transl., *J. Anal. Chem. USSR, 20*(9), 1018-1021 (1965).

Fluorine	Manganese
Chlorine	Technetium
Bromine	Rhenium
Iodine	

FLUORINE

Methods for the determination of fluorine include acid-base, redox, complexometric, and precipitation titrations [1,2].

Synoptic Survey

Classical Methods

Redox Titration. Fluoride is determined indirectly by precipitation with excess standard calcium acetate solution. The excess calcium is then converted to the oxalate which is titrated with standard permanganate [3].

Complexometric Titration. A hot neutral fluoride solution saturated with sodium chloride is titrated with a neutral aluminum chloride solution in the presence of methyl red. The AlF_6^{-3} ion is formed during the titration [4].

Precipitation Titration. Fluoride is titrated with standard thorium nitrate to a zirconium-alizarine lake end point [5].
Fluoride is precipitated as lead chlorofluoride which is dissolved in nitric acid and the chloride content determined by Volhard's method [6].

Contemporary Methods

Acid-Base Titration. Fluoride ion in hydrofluoric acid or in mixtures of sulfuric and fluosilicic acids is determined by potentiometric or high frequency titration with standard solution of sodium hydroxide [7].

Complexometric Titrations. Small amounts of fluoride in aqueous solution are titrated potentiometrically with tetraphenylantimony sulfate [8].

Cerium(IV) is complexed with fluoride, thereby lowering the cerium(IV)-cerium(III) redox potential. Unknown fluoride solutions are then added to a similar half-cell until the potential difference is zero [9].

Small amounts of fluoride are titrated photometrically with zirconium oxychloride using sodium 1,2,4-trihydroxyanthraquinone 3-sulfonate as the indicator [10].

Fluoride is titrated spectrophotometrically with standard lanthnum nitrate in aqueous alcohol solutions using 4-(2-pyrydylazo)resorcinol as the indicator [11].

Precipitation Titrations. Fluoride is precipitated as lead chlorofluoride with lead nitrate and known excess of hydrochloric acid. The excess unreacted chloride ion is then titrated potentiometrically [12].

Microamounts of fluoride are titrated spectrophotometrically with standard thorium nitrate solutions using sodium alizarine monosulfonate as the indicator [13]. The titration may also be carried out in dilute perchloric aicd solutions using methylthymol blue as the indicator [14].

In mixtures of fluoride, chloride and iodide, fluoride may be titrated conductometrically with standard calcium acetate. If desired, the titration of fluoride may be followed by a titration with mercuric nitrate in order to determine the chloride and iodide content of the sample [15].

Milligram samples of fluoride are titrated with europium(III) chloride or lanthanum(III) nitrate using a lanthanum fluoride membrane electrode (pF electrode) for end-point detection [16].

Samples containing fluoride may be determined by thermometric titration with thorium, cerium, aluminum, or calcium salts as titrants [17].

Fluoride is titrated with neodymium nitrate solutions. The end point is determined turbidimetrically [18].

Outline of Recommended Contemporary Methods for Fluorine

Acid-Base Titration

Fluoride ion in the form of hydrofluoric acid or in the presence of sulfuric and/or fluosilicic acids is determined by potentiometric titration with standard alcoholic sodium hydroxide solution using graphite and calomel electrodes. A single determination requires 30-40 minutes. The first break in the titration curve represents the sum of the sulfuric and fluosilicic acids, while the second inflection in the curve corresponds to the total amount of hydrofluoric acid, sulfuric acid, and fluosilicic acid present. A high frequency titration run in 60-70% aqueous acetone and using the same titrant may also be used to determine the above three acids. In this latter case, the first break in the titration curve corresponds to the sum of sulfuric and fluosilic acids, the second break shows the precipitation of sodium fluoride, and the third break corresponds to the sum of all three acids. The error in the potentiometric method is ± 2.4% for sulfuric acid and fluosilicic acid and ± 0.1% for hydrofluoric acid. In the high frequency titration, the error is ± 3.1 and ± 0.4% respectively [7].

Complexometric Titrations

Method 1. Aqueous samples of fluoride are determined by the use of 0.09 M tetraphenylantimony sulfate as an extractive titrant in the presence of chloroform. The relative error is $\pm$ 1.0% over a range of 1×10^{-3} M to 5×10^{-2} M fluoride. Fluoride is extracted as the ion pair $(C_6H_5)_4SbF$ into the organic phase. In the aqueous phase the fluoride ion activity is determined with a fluoride sensitive electrode. Phosphate, arsenate, arsenite, and sulfate do not interfere. Nitrate and perchlorate interfere but are readily removed by the addition of tetraphenylarsonium sulfate. Sulfite and nitrite interfere but are removed by oxidation with peroxide to sulfate and nitrate, the nitrate being removed as noted. Halides and thiocyanate interfere and are best removed by silver nitrate precipitation followed by extraction of excess nitrate with tetraphenylarsonium sulfate [8].

Method 2. Samples which are 0.05-0.5 M in fluoride are determined by null point potentiometry. The method is based on complexing cerium(IV) by fluoride and the lowering of the cerium(IV)-cerium(III) redox potential. Unknown fluoride is added to one Ce(IV)-Ce(III) half cell and standard fluoride solution to a similar half cell until the potential difference is zero. The method is applicable to the determination of fluoride in many binary and complex fluorides. Chloride, nitrate, and sulfate do not interfere. Interferences are shown by oxalate, phosphate and molybdate. The mean deviation is 0.07-0.20%. In the analysis of solutions which are 5×10^{-3} M in fluoride, the accuracy is 0.5% and 1% at 1×10^{-3} M [9].

Method 3. Samples containing not more than 6 mg fluoride in 80 ml of solution are titrated photometrically with 0.012 M zirconium oxychloride at pH 3.2 in the presence of a chloroacetate buffer. A 0.01% aqueous solution of sodium purpurin-3-sulfonate (sodium 1,2,4-trihydroxyanthraquinone 3-sulfonate) is used as the indicator. The titration is carried out at 550 nm. Titrant is added at the rate of 0.4 ml/min and the end point is obtained by extrapolation. Samples containing 6-25 mg fluoride are titrated with 0.05 M zirconium solutions at a pH adjusted to about 3 using a chloroacetate buffer and an external indicator such as 2,4-dinitrophenol. The adjustment of the pH must be made using an indicator since at this higher concentration of fluoride ion, the usual glass electrode of the pH meter is unsatisfactory. The standard deviation for samples containing 0.5 and 25 mg fluoride is $\pm$ 9 µg and 33 µg respectively [10].

Method 4. Samples containing not more than 9 mg fluoride in 25 ml aqueous alcohol are titrated photometrically at pH 6.5 (pyridine-pyridinium chloride buffer) with 0.017 M lanthanum nitrate using 4-(2-pyridylazo)resorcinol(PAR), as the indicator. The titration is carried out with stirring at 510 nm at a rate of 0.4 ml per minute of added titrant. The end point is obtained by extrapolation. Since the precipitate does not correspond exactly to LaF_3, an empirical factor is developed for the calculations. However, all variations from stoichiometry are less than 1%. The standard deviation for the above sample size is $\pm$ 5.0 µg. Cerium salts may also be used as the titrant [11].

Precipitation Titrations

Method 1. Fluoride is determined by precipitation with lead nitrate and a known amount of hydrochloric acid. The liquid phase is separated by filtration and the excess chloride in it is titrated potentiometrically at pH 4.5 with 0.32 N silver nitrate. A silver billet electrode and glass electrode combination is used with a standard pH meter for end-point detection. The unknown sample may contain up to 250 mg fluoride before the accuracy decreases appreciably. Most of the ions causing precipitation of lead do not interfere. Should such ions be present, additional lead nitrate solution is added. Iron and aluminum which form stable complexes with fluoride must be removed. Precision at the 200 mg level is $\pm$ 2.7 mg fluoride. The relative standard deviation for a sample of this size is $\pm$ 0.59% [12].

Method 2. Samples containing up to 200 µg of water soluble fluoride in 0.001 N nitric acid are titrated spectrophotometrically with 0.1 N aqueous thorium nitrate containing a portion of the indicator, sodium alizarin monosulfonate. The titration is carried out at 525 nm and the end point is obtained by extrapolation to zero absorbance. The intercept at zero absorbance is the volume of titrant at the end point. The fluoride content is then determined by reference to a calibration curve prepared by titration of 5-, 10-, and 20-ml portions of a standard fluoride solution. Using this procedure, fluoride is determined with a precision and accuracy of about $\pm$ 1.0% [13].

Method 3. Samples containing up to 10 mg of fluoride in no more than 15 ml of solution are acidified with 1 N perchloric acid to a pH of 3.35 $\pm$ 0.10. The solution is then buffered with a glycine-sodium perchlorate buffer and titrated with 0.02 M thorium nitrate solution to a methylthymol blue end point (dark blue). With this indicator, both precision and accuracy are significantly better than with other indicators such as alizarin red S or a mixed indicator such as xylenol orange-niagara sky-blue. Cations reacting with methylthymol blue interfere [14].

Method 4. Samples containing fluoride, chloride, and iodide ions in aqueous alcohol may be analyzed for fluoride content by titrating conductometrically with 0.1 N calcium acetate at a pH only slightly above 7. If desired, the iodide and chloride ions may be titrated in the same solution using 0.1 N mercuric nitrate [15].

Method 5. In the presence of 60 vol % ethanol, milligram samples of fluoride may be titrated with 0.037 M europium(III) chloride or 0.047 M lanthanum(III) nitrate using a lanthanum fluoride membrane electrode (pF electrode). Results are good. The electrode obeys the expected theoretical relation up to about the same pF value as in water solutions. In aqueous acid media, the upper limit of theoretical response (about pF 6.7) is greater than in neutral media (ca. 5.8) [16].

Method 6. Samples containing fluoride may be determined by

thermometric titration. Suitable titrants are standard solutions of
thorium, cerium, aluminum, and calcium ions. Although titration curves
with standard thorium solutions show the sharpest breaks, they are the
most sensitive to the presence of interfering ions. As a titrant,
aluminum has a high tolerance for sulfate and phosphate while aluminum
has a high tolerance for borate. By selection of the titrant and the
pH range, fluoride can be titrated in the presence of moderate to considerable amounts of sulfate, borate, phosphate, or silicate [17].

Method 7. Samples containing 1-20 mg of fluorine are titrated
with 0.2 M neodymium nitrate at pH 2.9 in a solution buffered with
chloroacetate. The end point is determined turbidimetrically. Variations in turbidity due to particle size change and coagulation during
the titration may be prevented by the addition of a few drops of 50%
v/v solution of polyethylene glycol 400. Some curvature in the titration curve at the beginning of the titration (induction period) may be
overcome by seeding with enough NdF_3 to produce a noticeable turbidity.
The average absolute error for the above sample sizes is 0.09 mg.

References

[1]. P. S. Elving, C. A. Horton, and H. W. Willard, eds., *Fluorine
      Chemistry*, Vol. 2, Chap. 3, Academic, New York, 1954.

[2]. C. A. Horton, in *Treatise on Analytical Chemistry* (I. M.
      Kolthoff and P. J. Elving, eds.), Part II, Vol. 7, Interscience,
      New York, 1961, p. 259.

[3]. W. W. Scott, *J. Ind. Eng. Chem.*, *16*, 703 (1924).

[4]. A. Kurtenacker and W. Jurenka, *Z. Anal. Chem.*, *82*, 210 (1930).

[5]. H. A. Williams, *Analyst*, *71*, 175 (1946).

[6]. N. H. Furman, ed., *Standard Methods of Chemical Analysis*, 6th
      Ed., Vol. 1, D. Van Nostrand, Princeton, N. J., 1962, pp. 450-
      451.

[7]. E. V. Bezrogova, *Zh. Analit. Khim.* *19*(12), 1498-1502 (1964).

[8]. J. B. Orenberg and M. D. Morris, *Anal. Chem.*, *39*(14), 1776-
      1780 (1967).

[9]. T. A. O'Donnell and D. F. Steward, *Anal. Chem.*, *33*, 337-341
      (1961).

[10]. C. Harzdorf, *Fresenius' Z. Anal. Chem.*, *232*(2), 172-180 (1967).

[11]. C. Harzdorf, *Fresenius' Z. Anal. Chem.*, *233*(5), 348-355 (1968).

[12]. T. M. Hogan and F. Tortori, *Anal. Chem.*, *39*(2), 221-223 (1967).

[13]. W. P. Pickhardt, *Anal. Chem.*, *34*, 863-864 (1962).

[14]. W. Selig, *Analyst*, *93*, 118-120 (1968).

[15]. L. M. Shtifman, A. A. Mityaeva, and V. T. Shemyatenkova, *Zavodsk. Lab.*, *31*(1), 39-40 (1965).

[16]. J. J. Lingane, *Anal. Chem.*, *40*(7), 935-937 (1968).

[17]. W. L. Everson and E. M. Ramirez, *Anal. Chem.*, *39*(14), 1771-1776 (1967).

[18]. J. E. Roberts, *Anal. Chem.*, *39*(14), 1884-1885 (1967).

CHLORINE

In this section devoted to chlorine, methods for the determination of elemental chlorine, chloride, hypochlorite, chlorite, chlorate, and perchlorate will be discussed.

Synoptic Survey of Elemental Chlorine Determinations

Methods for the determination of elemental chlorine include redox, complexometric, and precipitation titrations.

Classical Methods

Redox Titration. The iodine liberated by treatment of an iodide solution with free chlorine is titrated with standard thiosulfate [1].

Contemporary Methods

Complexometric Titration. Free chlorine is titrated amperometrically with standard phenylarsenic oxide [2].

Precipitation Titration. Free chlorine is reduced with bromide, and the chloride so formed is determined by differential potentiometry with standard silver nitrate [3].

Outline of Recommended Contemporary Methods for Elemental Chlorine

Complexometric Titration

Free chlorine is titrated amperometrically at pH 7 with 0.005 N phenylarsenic oxide using a platinum indicator electrode versus SCE. The titration is carried out until the diffusion current remains constant upon the addition of 0.1 ml of titrant. Chlorine in concentrations of 0.1-4.0 mg per liter can be determined with an accuracy of $\pm$ 0.04 mg/liter. The reaction, however, is not quantitative in the

presence of "Chemically bound effective chlorine." The latter, by
definition, is the total of all chlorinated products having oxidizing
and germicidal properties (e.g. the chloramines). When such are present,
a quantitative reaction between the free chlorine and the titrant,
phenylarsenic oxide, may be promoted by the addition of a small amount
of 0.1 N potassium iodide [2].

Precipitation Titration

Free chlorine is converted to chloride ion by reaction with aqueous potassium bromide. After removal of the excess bromine by an air
stream, the chloride is determined by differential potentiometry with
standard silver nitrate. A silver indicator electrode and a mercury(I)
sulfate reference electrode are used. In the range of 0.001 to 1.00%,
chlorine is determined with satisfactory accuracy. At 0.001% chlorine,
the relative standard deviation is 20% [3].

Synoptic Survey of Chloride Ion

Methods for the determination of chloride ion include complexometric, and precipitation titrations.

Classical Methods

Complexometric Titration. Chloride is titrated with standard
mercuric nitrate using a mixed indicator composed of diphenylcarbazone and bromophenyl blue [4].

Precipitation Titrations. Chloride is titrated in neutral solution with standard silver nitrate and using potassium chromate as the
indicator [5].
Chloride is precipitated in nitric acid solution with excess potassium thiocyanate. The excess is then back-titrated with standard
silver nitrate using ferric alum as the indicator [6,7].
Chloride is titrated with standard silver nitrate in the presence
of an absorption indicator such as fluorescein or dichlorofluorescein
[8,9].

Contemporary Methods

Precipitation Titrations. Low chloride concentrations may be
determined by constant current potentiometric titration with standard
silver nitrate [10].
Chloride and chloride-bromide mixtures are analyzed by amperometric titration with standard cadmium acetate in 0.2 M acetic acid
solution [11].

Outline of Recommended Contemporary Methods for Chloride

Precipitation Titrations

Method 1. Samples of chloride ion are determined by constant-

current potentiometric titration with 0.1 N silver nitrate using polarized platinum electrodes. The exact determination of low concentrations of chloride ion present in aqueous and non-aqueous media is feasible so that chloride ion in water and oil-soluble organic materials
can be determined without preliminary treatment. Tars and pitches may
be dissolved in alcohol and their chloride content determined with good
results [10].

    Method 2. Chlorides and bromides and their mixtures may be det-
ermined by amperometric titration with an acetic acid solution of cadmium acetate. A rotating amalgamated copper electrode and a copper
reference anode are used. The titration is carried out at -2.0 V. The
method is based on the successive precipitation of cadmium bromide and
cadmium chloride in anhydrous acetic acid. In mixtures, the first
break in the titration curve occurs at the equivalence point corresponding to titration of the bromide while the second corresponds to
the equivalence point for chloride. Mixtures of chlorides and brom-
ides varying from 20-80% chloride can be analyzed. Ammonium ions interfere. The maximum deviation from the mean for a series of 15 runs
is 0.1038; the standard deviation is 0.1865; and the maximum relative
percent deviation from the mean is ± 0.58% [11].

                    Synoptic Survey of Hypochlorite

    Methods for the determination of hypochlorite are limited pri-
marily to those involving redox titrimetry.

Classical Methods

    Redox Titration. Samples containing hypochlorite are treated
with potassium iodide and the liberated iodine is titrated with standard sodium arsenite to a starch end point [12]. The end point may
also be detected amperometrically [13].

Contemporary Methods

    Redox Titrations. Hypochlorite is titrated potentiometrically
with standard ferrous sulfate [14].
    Hypochlorite is determined by adding it to a solution containing
a known excess of arsenite and iodide. The excess arsenite is then
determined by coulometric generation of iodine [15].
    Solutions containing hypochlorite are treated with an excess of
thallium(I) sulfate. The thallium(III) so formed is titrated with
standard ascorbic acid [16].

    Outline of Recommended Contemporary Methods for Hypochlorite

Redox Titrations

    Method 1. Samples containing from 1-40 mg hypochlorite are
titrated potentiometrically with 0.01 N or 0.1 N iron(II) sulfate

using a platinum indicator electrode versus SCE. The potential jump
at the end point is about 510 mV at pH 4.9-5.5. The mean error is $\pm$
0.3%. Sulfate, chlorate, chloride, and nitrate do not interfere. Bromide, iodide, and phosphate must be removed prior to the titration.
Chlorite is titrated simultaneously. The method is recommended for
the determination of hypochlorite in chlorinated lime [14].

Method 2. Samples containing hypochlorite are determined by an
indirect coulometric titration. This titration is based on two simultaneous reactions of hypochlorite.
    In weakly basic solution, hypochlorite reacts with arsenite
according to the equation:

$$ClO^{-1} + As^{+3} = Cl^{-1} + As^{+5} \qquad (A)$$

Hypochlorite also reacts with iodide:

$$ClO^{-1} + 2 I^{-1} = I_2 + Cl^{-1} \qquad (B)$$

And the iodine formed also reacts with arsenite:

$$I_2 + As^{+3} = 2 I^{-1} + As^{+5} \qquad (C)$$

By introducing samples containing hypochlorite into a solution containing a known excess of arsenite and iodide, hypochlorite reacts
either directly, equation A, with arsenite, or indirectly via equations B and C. In either case, hypochlorite is consumed and measured.
The excess arsenite then is determined by the coulometric generation
in a slightly basic solution (pH 8.0) of iodine from the iodide present.
The end-point detection is biamperometric. The determination may be
carried out rapidly, and the precision is 1-2 parts per thousand [15].

Method 3. Hypochlorite is determined by the addition of an excess of thallium(I) sulfate to the unknown hypochlorite solution. The
thallium(III) formed by the oxidation is precipitated as the hydroxide.
The pH of the solution is then adjusted to pH 4.1 with an acetate buffer, and the thallium(III) is titrated with standard ascorbic acid to
a variamine blue end point. The standard deviations for 0.1 N and
0.01 N NaOCl are $\pm$ 0.07 and $\pm$ 0.06 [16].

Synoptic Survey of Chlorite Determinations

    The only satisfactory reported methods for the determination of
chlorite involve redox titrimetry.

Classical Methods

    Redox Titration. Chlorite ion oxidizes iodide in acetic acid
solution to iodine which is titrated with standard thiosulfate [17].

Contemporary Methods

    Redox Titration. Samples of chlorite are treated with excess ferrocyanide ion. The ferricyanide formed is then titrated with standard ascorbic acid [18].

Outline of a Recommended Contemporary Method for Chlorite

Redox Titration

Samples containing chlorite are treated with an excess of ferrocyanide.

$$4 \text{ H}^+ + \text{ClO}_2^{-1} + 4[\text{Fe(CN)}]_6^{-4} = \text{Cl}^{-1} + 4[\text{Fe(CN)}]_6^{-3} + 2 \text{ H}_2\text{O}$$

The ferricyanide formed is titrated with standard ascorbic acid solution in a buffered bicarbonate solution to a 2,6-dichloroindophenol end point. For samples containing 20 mg chlorite, the standard deviation is $\pm$ 0.3% [18].

Synoptic Survey of Chlorate Determinations

Methods for the determination of chlorate involve redox titration.

Classical Methods

Redox Titration. Chlorate is reduced by a known excess of ferrous sulfate. The unreacted ferrous ion is back-titrated with standard potassium permanganate or potassium dichromate [19].

Contemporary Methods

Redox Titration. Samples containing chlorate ion are reduced with nitrite or disulfide and the chloride ion formed is titrated with standard silver nitrate using dichlorofluorescein or potassium dichromate as the indicator [20].

Outline of Recommended Contemporary Methods for Chlorate Ion

Redox Titration

Samples containing 100-500 mg chlorate are treated with 1 M sodium nitrite and the mixture acidified with dilute nitric acid. After boiling and cooling, the chloride content is determined by titration with 0.1 M silver nitrate using dichlorofluorescein as the indicator. Alternately, a Mohr titration may be used. The reduction of the chlorate to chloride may also be carried out by boiling for several minutes with 10% sodium disulfide whereupon the solution is acidified with 3 N nitric acid and the boiling is continued until no more sulfur dioxide is evolved. All anions which give sparingly soluble silver salts and all cations which give sparingly soluble chromates, chlorides, or sulfates must be absent. The error in the determination is about $\pm$ 0.2% [20].

Synoptic Survey of Perchlorate Determinations

Recommended methods for the determination of perchlorate include redox and precipitation titrations.

Classical Methods

Precipitation Titration. Perchlorate is reduced with zinc amalgam to chloride in the presence of molybdate as a catalyst. The chloride formed is determined by a Volhard titration [21].

Perchlorate is determined by fusion with sodium peroxide followed by titration of the chloride formed [22].

Contemporary Methods

Redox Titrations. Samples of perchlorate are reduced with a known excess of titanous chloride. The unreacted titanous ion is then back-titrated with standard ferric ammonium sulfate [23].

Perchlorate is determined by reduction with a known excess of standard ferrous sulfate. The excess ferrous ion is then back-titrated with standard potassium permanganate [24].

Precipitation Titrations. Perchlorate is titrated conductometrically with tetraphenylarsonium chloride [25]. A similar titration may be carried out potentiometrically [26].

Perchlorate is determined by reduction to the chloride with solid vanadium(III) sulfate. The chloride formed is then titrated potentiometrically with silver nitrate [27].

Outline of Recommended Contemporary Methods for Perchlorate

Redox Titrations

Method 1. Perchlorate is reduced by 10 minutes boiling with a 100 mole % excess of 0.3 N titanium(III) chloride. The excess is then back-titrated with standard ferric ammonium sulfate. The reaction involved in the reduction is:

$$ClO_4^- + 8\ Ti^{+3} + 4\ H_2O = Cl^- + 8\ TiO^{+2} + 8\ H^+$$

The back-titration proceeds according to the equation:

$$Ti^{+3} + Fe^{+3} + H_2O = TiO^{+2} + Fe^{+2} + 2\ H^+$$

Osmium tetroxide in sulfuric acid is used as the catalyst. This procedure has been used for the assay of rocket grade ammonium perchlorate. The maximum standard deviation found is 0.19% [23].

Method 2. Samples containing perchlorate are reduced to chloride by heating the sample with an excess of standard iron(II) sulfate in strong sulfuric acid solutions at 150-155° for 15 minutes. The unreacted iron(II) sulfate is then back-titrated with standard potassium permanganate [24].

Precipitation Titrations

Method 1. Samples containing 1-1.5 mmoles of perchlorate in 200 ml of water are titrated conductometrically with 0.05 M tetraphenylarsonium chloride at pH 7. A dip-type platinized platinum electrode having a cell-constant of 0.7 cm^{-1} is used for the conductance measurements. The sample is titrated at the rate of about 2 ml per minute. By use of the above sample size and the recommended concentrations of titrant and unknown, it is possible to keep the conductance of the solution fairly constant before the end point. Extrapolation of the linear segments of the titration curves include angles of less than 120° and thus the end-point intersection is readily seen. There are no interferences from iron(III), sulfate, iodate, fluoride, molybdate, phosphate, chloride, chlorate, bromide, bromate, nitrate, or chromate. Iodide and permanganate interfere. The precision of the method expressed at the 95% confidence limit is ± 0.16% [25].

Method 2. Samples of perchlorate 1.5-2.0 mmoles in 75 ml of aqueous solution are titrated potentiometrically with 0.05 M tetraphenylarsonium chloride at pH 4-7. The titration is carried out using a perchlorate ion specific electrode and a double-junction calomel electrode. The method is simple and rapid. No interferences are met from chloride, chlorate, bromide, bromate, fluoride, sulfate, nitrate, or chromate. Dichromate, iodide, iodate, and permanganate interfere. The precision of the method at the 95% confidence limit is ± 0.16% [26].

Method 3. Samples of perchlorate are reduced to chloride by refluxing for 5-10 minutes in 7-8 M sulfuric acid with air-stable solid vanadium(III) sulfate. Osmium tetroxide dissolved in sulfuric acid is used as the catalyst. The chloride formed is determined by potentiometric titration with standard silver nitrate. The accuracy is 0.1-0.2% for samples containing 0.01-3.0 mmoles of perchlorate [27].

References

[1]. N. H. Furman, ed., *Standard Methods of Chemical Analysis*, 6th Ed., Vol. 1, D. Van Nostrand, Princeton, N. J., 1962, p. 333.

[2]. U. Haesselbarth, *Fresenius' Z. Anal. Chem.*, *234*(11), 22-37 (1968).

[3]. J. Meyer, *Fresenius' Z. Anal. Chem.*, *212*(2), 292-302 (1965).

[4]. F. E. Clarke, *Anal. Chem.*, *22*, 5531-5535 (1950).

[5]. W. G. Young, *Analyst*, *18*, 125 (1893).

[6]. E. Swift, G. M. Arcand, F. Lutwack, and D. J. Meier, *Anal. Chem.*, *22*, 30 (1950).

[7]. S. Stanton, *Anal. Chem.*, *23*, 1351 (1951).

[8]. K. Fajans and H. Wolf, *Z. Anorg. Allgem. Chem.*, *137*, 33 (1924).

[9].   I. M. Kolthoff, W. M. Lauer, and C. J. Sunde, *J. Am. Chem. Soc.*, *21*, 3273 (1929).

[10]. R. W. Freedman, *Anal. Chem.*, *31*(2), 214 (1959).

[11]. A. P. Kreshkov, V. A. Bork, and K. S. Salnikova, *Zh. Anal. Khim.* *22*(9), 1392-1397 (1967). Eng. transl., *J. Anal. Chem. USSR*, *22*(9), 1173-1177 (1967).

[12]. I. M. Kolthoff, *Rec. Trav. Chim.*, *41*, 740 (1922).

[13]. S. S. Listek, *Anal. Chem.*, *20*, 639 (1948).

[14]. J. Vulterin and J. Hovorka, *Collect. Czech. Chem. Commun.*, *32* (11), 4063-4069 (1967).

[15]. A. F. Krivis and E. S. Gazda, *Anal. Chem.*, *39*(2), 226-227 (1967).

[16]. L. Erdey and K. Vigh, *Talanta*, *10*, 439-444 (1963).

[17]. W. Wagner, C. J. Hull, and G. E. Markle, *Advanced Analytical Chemistry*, Reinhold, New York, 1956, p. 232.

[18]. L. Erdey and G. Sevhla, *Z. Anal. Chem.*, *167*, 164-172 (1959).

[19]. W. Wagner, C. J. Hull, and G. E. Markle, *Advanced Analytical Chemistry*, Reinhold, New York, 1956, p. 232.

[20]. C. J. Coetzee, *Fresenius' Z. Anal. Chem.*, *234*(4), 245-247 (1968).

[21]. G. P. Haight, Jr., *Anal. Chem.*, *25*, 642-643 (1953).

[22]. N. L. Crump and N. C. Johnson, *Anal. Chem.*, *27*(16), 1007-1008 (1955).

[23]. E. A. Burns and R. F. Muraca, *Anal. Chem.*, *32*, 1316-1319 (1960).

[24]. G. Aravamudan and V. Khrishnan, *Talanta*, *13*(3), 519-522 (1966).

[25]. R. J. Baczuk and W. T. Bolleter, *Anal. Chem.*, *39*(11), 93-95 (1967).

[26]. R. J. Baczuk and R. J. Du Bois, *Anal. Chem.*, *40*(4), 685-689 (1968).

[27]. D. A. Zatko and B. Kratochvil, *Anal. Chem.*, *37*(2), 1560-1562 (1965).

BROMINE

In this section, the titrimetric methods for the determination
of elemental bromine, bromide, hypobromite, bromite, and bromate will

be discussed. However, since the analytical chemistry of bromine is
similar to that of chlorine, many of the methods applicable to chlorine
and its compounds may also be applied to bromine.

Synoptic Survey of Elemental Bromine Determinations

       Methods for the determination of elemental bromine involve pri-
marily redox titration.

Classical Methods

       Redox Titrations. Samples containing bromine are treated with a
potassium iodide solution. The liberated iodine is then titrated with
standard thiosulfate [1].
       Free bromine may be titrated potentiometrically with hydrazine
[2].
       Bromine is oxidized in an acidic solution in the presence of
cyanide to BrCN which in turn is treated with potassium iodide. The
liberated iodine is then titrated with standard thiosulfate [3].

Contemporary Methods

       No recent advances in the determination of free bromine have been
reported.

Synoptic Survey of Bromide Ion Determinations

       The only reported and useful method for the determination of
bromide ion involves precipitation titration.

Classical Methods

       Precipitation Titrations. Samples containing bromide are treated
with excess silver nitrate and the unreacted silver ion is back-titrated
with standard ammonium thiocyanate using ferric alum as the indicator
(Volhard's method) [4].
       Bromide is precipitated by the addition of excess standard mer-
curic nitrate. The excess is back-titrated with standard potassium
bromide in the presence of ferroin as the indicator [5].

Contemporary Methods

       No recent developments have been reported in titrimetric pro-
cedures for bromide ion.

Synoptic Survey of Hypobromite and Bromite Determinations

       Methods for the determination of hypobromite and bromite involve,
at the present time, redox titration.

Classical Methods

Redox Titrations. Samples containing hypobromite are allowed to react with excess potassium iodide solution, and the iodine formed is titrated with standard thiosulfate [6].

Hypobromite and hypochlorite when present in the same solution may be differentiated by titrating one aliquot with standard arsenite to measure both oxyacids. A second aliquot is treated with phenol to remove the hypobromite and a second titration with arsenite is then performed to determine the hypochlorite [7].

Contemporary Methods

Redox Titrations. An excess of thallium(I) sulfate is added to the hypobromite solution and the thallium(III) ion formed is titrated reductometrically with standard ascorbic acid [8,10,11].

Samples containing both hypobromite and bromite are titrated potentiometrically with standard arsenite. Titration at a low pH gives the hypobromite value. Titration at pH 9.7-10.0 gives the sum of both constituents [9].

Hypobromite is titrated potentiometrically in the presence of bromite with standard potassium ferrocyanide. If the titration is carried out at pH 9.7-9.9, the total amount of both ions is measured [9].

Samples of hypobromite in alkaline solutions are titrated with standard hydrazine solutions using Bordeaux B as the indicator [12].

Hypobromite samples are treated with an alkaline solution of vanadium(IV). The vanadium(V) formed is titrated with standard ferrous sulfate using phenanthroline as the indicator [13].

Outline of Recommended Contemporary Methods for Hypobromite and Bromite

Redox Titrations

Method 1. An excess of 0.1 N thallium(I) sulfate is added to the hypobromite solution and the thallium(III) ions which have been formed in an equivalent amount are titrated reductometrically at pH 4.1 (acetate buffer) with 0.1 N ascorbic acid. Variamine blue is used as the indicator. The standard deviation for 0.1 N or 0.01 N NaOBr is ± 0.06 and 0.08, respectively. Chlorite ion does not interfere if its concentration is less than that of hypochlorite [8]. Bromate in large amounts interferes [10,11].

Method 2. Hypobromite and bromite are determined successively in the same sample by titration with 0.01 N sodium arsenite solution. A platinum electrode versus a calomel electrode is used in the titrations for end point detection. The first titration is carried out at a pH 8-9 and the end point corresponds to the amount of hypobromite present. The pH is then raised to 9.7-9.9, osmium tetroxide is added as a catalyst, and the titration is continued to the second end point which measures the sum of both constituents. The error is less than 2% and the method has been successfully used for the quantitative studies on the radiolysis products from magnesium, calcium, and cadmium bromates [9].

Method 3. Due to the slowness of the reaction between bromite
and ferrocyanide, hypobromite can be titrated potentiometrically in
the presence of bromite with 0.01 N potassium ferrocyanide. As noted
above, the sum of the constituents is found by a potentiometric titration at pH 9.7-9.9 with standard arsenite. This method may also be
used for the determination of small amounts of hypobromite and bromite
in solutions of irradiated alkali bromates. In such solutions, the
total amount of hypobromite and bromite formed at a dose of 50 Mrads is
approximately 1 mg of bromite per 200 mg of bromate. In routine analysis, such amounts can be determined with a relative error of 2-3%
[9].

Method 4. Alkaline solutions of hypobromite are titrated with
standard hydrazine solutions. Bordeaux B is used as the indicator but
it must be added to the solution being titrated only 0.1 ml before the
equivalence point. Bromite ion interferes [12].

Method 5. Samples containing 1.0×10^{-4} - 2.5×10^{-4} moles hypochlorite or hypobromite are treated with a known excess of a deoxygenated basic 0.05 M vanadium(IV) sulfate solution. After acidification
with sulfuric acid, the vanadium(V) is titrated with standard ferrous
sulfate using phenthroline as the indicator. For hypobromite, the
percent error at a maximum is + 0.37%. Cerium(IV), chromium(VI), and
other ions reduced by iron(II) interfere [13].

Synoptic Survey of Bromate Determinations

Methods for the determination of bromate include redox and complexometric titrations.

Classical Methods

Redox Titrations. Samples of bromate are reduced with a known
excess of arsenious acid. The excess is back titrated with standard
iodine solution [14].
Bromate is reduced by iodide ion in an acidic solution in the
presence of sodium molybdate as a catalyst. The iodine formed is titrated with standard sodium thiosulfate [15,16].

Contemporary Methods

Redox Titrations. In a mixture containing hypobromite, hypobromous acid, bromide, and bromate, the bromate concentration is determined by the difference of the two titrations. The first, a potentiometric titration, gives the total hypobromous acid and hypobromite
content. The second, an iodometric titration carried out on a separate aliquot, measures the sum of the hypobromous acid, hypobromite,
and bromate [17].
Bromate is determined by potentiometric titration with ferrous
ammonium sulfate [18].

Complexometric Titration. In a mixture of bromide and bromate,
the bromide content is determined by precipitation as silver bromide

which is dissolved in potassium tetracyanonickelate, and the liberated nickel ion is then titrated with standard EDTA. In a separate aliquot, the bromate is reduced to bromide with arsenic(III) ion, and the total bromide is determined in the same manner [19].

Outline of Recommended Contemporary Methods for Bromate

Redox Titrations

Method 1. Samples containing hypobromite, hypobromous acid, bromide, and bromate are analyzed by a combination of potentiometric, iodometric, and argentometric titrations. In one aliquot, the sum of the hypobromous acid and hypobromite is determined by a potentiometric titration with a standard arsenic(III) solution. In a second aliquot, the total hypobromous acid, hypobromite and bromate content is determined iodometrically. The difference in these two titrations allows one to calculate the bromate concentration. In a third aliquot, the bromide content is determined after prior treatment with hydrogen peroxide. Such treatment reduces all of the oxyanions to bromide which, after acidification with nitric acid to pH 5-7, is titrated potentiometrically with standard silver nitrate using the silver bromide electrode [17].

Method 2. Samples containing 0.30-0.45 mg bromate per ten milliliters of solution are added to an excess of 85% phosphoric acid, and the resulting solution is titrated potentiometrically with 0.01-0.1 N ferrous ammonium sulfate using a platinum electrode versus SCE. The first break in the titration curve occurs in 10.5 M phosphoric acid at about 500 mV versus SCE. Here the line slope is 3500, and the break corresponds to the reduction of the bromate to free bromine. The second break occurs at about 500 mV versus SCE and has a line slope of 13,500. This break occurs at the equivalence point in the reaction of the free bromine with the iron(II).

$$Br_2 + 2\ Fe^{+2} = 2\ Br^- + 2\ Fe^{+3}$$

If one carries out the titration in 9-13 M phosphoric acid, then both breaks may be used for accurate determinations. In low concentrations of phosphoric acid, however, only the first break may be evaluated since loss of bromine readily occurs. The standard deviation is $\pm$ 0.20% relative. Sulfate, phosphate, or bromide do not interfere. Chloride does not interfere up to a molar ratio of 1:1. Nitrate, iodide, chlorate, iodate, and periodate are also reduced and, therefore, interfere in the determination [18].

Complexometric Titration

In a mixture of bromide and bromate, the bromide is precipitated in an acidic solution as silver bromide. The silver bromide is dissolved in an ammoniacal solution of potassium tetracyanonickelate. The equivalent amount of nickel ions which are liberated are then titrated with standard EDTA. In another aliquot of the sample, the bromate is reduced to bromide ion by treatment with arsenic(III), and the total bromide content is then determined as previously described [19].

References

[1]. N. H. Furman, ed., *Standard Methods of Chemical Analysis*, 6th
      Ed., Vol. 1, D. Van Nostrand, Princeton, N. J., 1962, p. 242.

[2]. W. R. Crowell, *J. Am. Chem. Soc.*, *58*, 1324-1328 (1932).

[3]. R. Lang, *Z. Anorg. Chem.*, *144*, 75-84 (1925).

[4]. N. H. Furman, ed., *Standard Methods of Chemical Analysis*, 6th
      Ed., Vol. 1, D. Van Nostrand, Princeton, N. J., 1962, pp. 242-
      243.

[5]. M. E. Hall and G. M. Smith, *Anal. Chem.*, *23*, 1181 (1951).

[6]. I. M. Kolthoff, *Rec. Trav. Chim.*, *41*, 740 (1922).

[7]. L. Farkas and M. Lewin, *Anal. Chem.*, *19*, 662-664 (1947).

[8]. L. Erdey and K. Vigh, *Talanta*, *10*, 439-444 (1963).

[9]. T. Andersen and H. E. L. Madsen, *Anal. Chem.*, *37*(1), 49-51
      (1965).

[10]. M. H. Hashmi and A. A. Ayaz, *Analyst*, *88*, 147 (1963).

[11]. M. H. Hashmi, A. Rashid, A. A. Ayaz, and N. A. Chughatai, *Anal.
      Chem.*, *38*(3), 507-508 (1966).

[12]. M. H. Hashmi, A. A. Ayaz, and A. S. Maqbool, *Z. Anal. Chem.*,
      211(6), 417-419 (1965).

[13]. J. P. Tandon and K. L. Chawia, *Bull. Chem. Soc. Jap. 39*(11)
      254-255 (1966).

[14]. F. A. Gooch and J. C. Blake, *Am. J. Sci.*, *14*, 285 (1902).

[15]. I. M. Kolthoff, *Z. Anal. Chem.*, *60*, 348 (1921).

[16]. F. Taradoire, *Bull. Soc. Chim. France 4*(5), 1759-1771 (1937).

[17]. I. E. Flix, G. I. Pusenok, and M. K. Bynyaeva, *Tr. Leningr.
      Tekhnol. Inst. Tsellulozn.-Bumazhn. Prom.*, *11*, 111-117 (1962).

[18]. J. Vulterin, *Collect. Czech. Chem. Commun.*, *32*(9), 3349-3357
      (1967).

[19]. A. de Sousa, *Z. Anal. Chem.*, *174*, 337-339 (1960).

IODINE

In this section, selected titrimetric methods for the determination of elemental iodine, iodide, iodate, and periodate will be discussed. Many of the methods applicable to chlorine and its compounds may also be applied to iodine.

Synoptic Survey of Elemental Iodine Determinations

Methods for the determination of elemental iodine involve primarily redox titration.

Classical Methods

Redox Titrations. Iodine in solution is measured by titration against standard arsenite or standard thiosulfate in neutral solution using starch as the indicator [1,2]. The titration may also be carried out using the dead-stop procedure by adding the thiosulfate into the iodine solution in the presence of two inert electrodes [3].

Samples containing iodine are titrated in dilute acid solution in the presence of an acetate buffer with standard ascorbic acid solutions using variamin blue as the indicator [4].

Contemporary Methods

Redox Titration. A recent modification of the method above using ascorbic acid as the titrant involves a change in indicator from variamin blue to 2-hydroxy-4-amino-4'-methoxydiphenylamine (2-hydroxy-variamin blue). Bromine, iodate, and bromate may also be determined using this method [5,6].

Outline of Recommended Contemporary Methods for Elemental Iodine

Redox Titration

Samples containing 62-254 mg iodine are titrated with 0.1 N ascorbic acid in the presence of an acetate buffer. A 2-hydroxyvariamin blue solution is used as the indicator. The solution of the indicator is prepared by mixing 1 part of the 2-hydroxyvariamin blue with 500 parts of sodium chloride. A portion of this mixture weighing 0.3-0.9 gm is then used for each titration. With this amount of indicator, the end-volume on the titration should be about 100 milliliters. From twelve parallel determinations, the calculated standard deviation amounts to $\pm$ 0.06%.

The determination of iodine may also be carried out indirectly by adding to the iodine sample an excess of 0.1 N potassium ferrocyanate which is oxidized by the iodine to the ferricyanate. The unreacted ferrocyanate ion is then titrated with ascorbic acid as above using 2-hydroxyvariamin blue as the indicator in a solution buffered with bicarbonate. From twelve parallel determinations, the calculated standard deviation amounts to $\pm$ 0.11% [5,6].

Synoptic Survey of Iodide Determinations

Methods for the determination of iodide include redox and precipitation titrimetry.

Classical Methods

Redox Titration. Iodide is oxidized with iodate, and the liberated iodine is titrated with standard thiosulfate [7,8]. The oxidation
of the iodide may also be carried out using iron(III) solutions [9] or
nitrous acid [10].

Precipitation Titrations. Samples containing iodide are treated
with excess silver nitrate and the unreacted silver ion is back titrated with standard thiocyanate in Volhard's method with ferric alum as
the indicator [11].
Iodide is titrated in neutral or slightly acidic solutions with
standard silver nitrate using eosin as an adsorption indicator [12].
Other adsorption indicators that may be employed are: rose bengal
[13], p-ethoxychrysoidine [14], or red acid 6B [15].

Contemporary Methods

Redox Titrations. Small amounts of iodide in the range of 5-100
ppm are determined directly in the presence of chloride by null-point
potentiometric method [16].
Samples containing both iodide and thiosulfate are analyzed for
each constituent by a double titration involving a titration with
standard iodine to determine the thiosulfate followed by oxidation of
the iodide and a subsequent titration with standard thiosulfate [17].
Iodide samples are analyzed by titration with cerium(IV) sulfate
solutions using a dead-stop technique [18].
Iodide is determined by potentiometric titration with sodium
peroxymolybdate [19].

Outline of Recommended Contemporary Methods for Iodide

Redox Titrations

Method 1. Samples containing micro amounts (5-100 ppm) iodide
in the presence of large concentrations of chloride are analyzed by
the precision null-point method. This procedure is based on rapidly
changing the iodide concentration of an unknown sample until it has
the same concentration as a constant reference iodide solution. A
potentiometric system is used to indicate the null-point together with
a pair of silver-silver iodide electrodes sensitive to the iodide concentration. Samples in the above range are determined with an average
deviation of less than 0.3 ppm and with relative errors of about 0.5%
at 100 ppm and 5% at 10 ppm even when the sodium chloride to iodide
weight ratios are as high as 200,000 to 1 [16].

Method 2. Samples containing 1-5 mg each of iodide and thiosulfate in the same solution are determined by the dead-stop method.

The thiosulfate is first titrated with 0.05 N iodine solution. The total iodide (from the titration and that present in the original sample) is oxidized by the addition of hypochlorite (containing 90 gm of active chlorine per liter) to iodate. The solution is acidified with sulfuric acid, boiled to remove any excess hypochlorite, cooled, and the iodate is determined iodometrically. The iodide concentration of the original sample is obtained by difference. Bromide, thiocyanate, chloride, and phosphate do not interfere [17].

Method 3. Iodide in samples containing 0.1 meq iodide are determined by titration in 2 N sulfuric acid with 0.01 N cerium(IV) sulfate. The end point is determined by the usual dead-stop technique. The potential applied to the system is about 25 mV and titration is continued to a current minimum. Using this method, iodide may be determined in the presence of tenfold excess bromide [18].

Method 4. Samples containing about 0.25 meq iodide in 100 ml of 5 N sulfuric acid are titrated, after the addition of a few drops of 0.05 N potassium iodide as a catalyst, with sodium peroxymolybdate. The titration is carried out potentiometrically using a platinum electrode versus SCE. The mean error in five determinations for a 19 mg sample of iodide is 0.63%. Arsenite may also be determined by this method [19].

Synoptic Survey of Iodate Determinations

Iodate is most generally determined by redox titration.

Classical Methods

Redox Titration. The iodine formed by the reaction of samples containing iodate with iodide in acid solution is titrated with standard thiosulfate [20]. The end point in the reaction may also be determined potentiometrically [21]. The reduction of the iodate may also be carried out with hydrazine [22], stannous chloride [23], or ascorbic acid [24].

Contemporary Methods

Redox Titrations. Iodate is titrated potentiometrically in phosphoric acid solution with standard ferrous ammonium sulfate. The reaction is catalyzed by the addition of osmium tetroxide [25].

Samples of iodate are allowed to react with excess ferrocyanide. The excess is back-titrated potentiometrically with ascorbic acid [26].

Iodate is measured in the presence of periodate by masking the periodate with molybdate followed by titration with thiosulfate [27].

Samples of iodate are treated with excess iodide and the unreacted iodide is titrated potentiometrically with mercury(II) nitrate [28].

Outline of Recommended Contemporary Methods for Iodate

Redox Titrations

Method 1. Samples containing 0.3–80.0 mg iodate in 30 ml water
are acidified with 85% phosphoric acid. After adding 0.2 ml of 0.01 N
osmium tetroxide, the sample is titrated potentiometrically with 0.001-
0.01 N ferrous ammonium sulfate. The mean relative deviation is $\pm$
0.15%. The slope of the curve at the inflection point is 7000 with a
potential of 730 mV versus SCE. A large excess of phosphate, sulfate,
tetraborate, a threefold excess of chloride, a thirtyfold excess of
fluoride or a seventyfold excess of nitrate does not interfere. Bromide and iodide interfere. The method is based on the reaction:

$$2 \ IO_3^- \ + \ 10 \ Fe^{+2} \ + \ 12 \ H^+ \ = \ I_2 \ + \ 10 \ Fe^{+3} \ + \ 7 \ H_2O$$

In 0.1-12 M phosphoric acid, the redox potential of the system
IO_3^-/I_2 is 0.97-1.17 V versus standard hydrogen electrode [25].

Method 2. Samples containing iodate are reduced with excess
standard potassium ferrocyanide followed by back titration of the unreacted ferrocyanide with standard ascorbic acid. Titrations are best
carried out with an automatic titrimeter which records the potentiometric titration curve. For samples containing 0.1-1.5 mg iodate, the
relative standard deviation is 1.19%, and the relative error for samples
containing 0.04 meq of iodate is 1.33%. The method may also be applied
to the titrimetry of ferricyanide, ferrocyanide, dichromate, permanganate, and bromate [26].

Method 3. Samples containing 4-10 x 10^{-5} moles periodate and 2.6-
7.7 x 10^{-5} moles iodate are determined, after buffering with a chloro-
acetic acid buffer (pH 2.9-3.0) and masking the periodate with molybdate, by an iodometric titration. This titration with 4.0 x 10^{-2} M
thiosulfate is continued until the solution becomes straw-colored.
A little thyoden is then added and the titration is continued to the
disappearance of the blue color. The periodate is then demasked by
the addition of oxalic acid, and the liberated iodine is again titrated with thiosulfate. The pH of the solution after both titrations is
about 1.7. Precision and accuracy are excellent [27].

Method 4. A known volume of the unknown iodate solution is added
to an excess but known amount of 0.2 N potassium iodide. The solution
is made about 2 N with respect to sulfuric acid, and the unreacted potassium iodide is titrated potentiometrically with mercury(II) nitrate
using a silver amalgam indicator electrode versus SCE. Relatively large
amounts of iron(III), copper(II), or arsenic(V) do not interfere. An
alternate procedure involves adding the unknown iodate solution to 2 N
sulfuric acid followed by the addition of solid sodium sulfite. The
iodide formed by this reduction is then titrated as above. This method
is also applicable to the determination of permanganate, chromate, dichromate, bromate, periodate, and cerium(IV). The procedure is simple,
rapid, shows high accuracy and precision and may be applied to the
analysis of a wide concentration range of the above ions [28].

Synoptic Survey of Periodate Determinations

Methods applicable to the determination of periodate involve primarily redox titrimetry.

Classical Methods

Redox Titrations. The iodine formed by the reaction of periodate and iodide in acidic solutions is titrated with standard sodium thiosulfate [29].

Periodate is titrated with 0.1 N arsenic trioxide in dilute sulfuric acid medium with ferroin as the indicator. The reaction is catalyzed by ruthenium salts [30].

Periodate may be quantitatively separated from iodate by precipitation with zinc acetate. After removal of the precipitate, it is dissolved and titrated with thiosulfate. The iodate may be determined iodometrically in the filtrate [31].

Contemporary Methods

Redox Titrations. Periodate is titrated potentiometrically in phosphoric acid medium with standard ferrous ammonium sulfate. Osmium tetroxide is used to catalyze the reaction [32].

Samples containing periodate ion which is in excess after carbohydrate oxidation are determined iodometrically at pH 7.3 in a solution buffered with borate [33].

Outline of Recommended Contemporary Methods for Periodate

Redox Titrations

Method 1. Samples containing 0.09–70.0 mg periodate in 30 ml of water are treated with 20 ml of 85% phosphoric acid, 0.2 ml of 0.01 M osmium tetroxide and then titrated potentiometrically with 0.001–0.01 M ferrous ammonium sulfate. A platinum indicating electrode is used. The slope of the curve at the inflection point in 3–6 M phosphoric acid is 15,000 at a potential of 750 mV versus SCE. Fluoride ion in fourteen fold amounts, or chloride in twenty-fivefold amounts does not interfere. Likewise a large excess of phosphate, tetraborate, or nitrate does not interfere. Bromide and iodide interfere. The mean relative deviation is $\pm$ 0.26%. The method is based on the reduction of the periodate according to the following equation and the subsequent titration of the iodine in the usual manner.

$$2 \ IO_4^- + 14 \ Fe^{+2} + 16 \ H^+ = I_2 + 8 \ H_2O + 14 \ Fe^{+3}$$

Using this method 30 γ of periodate in 30 ml of solution can be determined by titration with 0.001 M ferrous ammonium sulfate [32].

Method 2. Samples of the solution remaining after periodate oxidation of oligosaccharides or deoxysaccharides and which still contain unreacted periodate may readily be analyzed for their periodate content. The sample is neutralized with bicarbonate and the pH is adjusted to pH 7.3 with a borate buffer. An excess of KI is added and the liberated iodine is determined by titration with thiosulfate [33].

References

[1]. N. H. Furman, ed., *Standard Methods of Chemical Analysis*, 6th Ed., Vol. 1, D. Van Nostrand, Princeton, N. J., 1962, pp. 517-518.

[2]. C. L. Wilson and D. W. Wilson, eds., *Comprehensive Analytical Chemistry*, Vol. 1c, Elsevier, New York, 1962, p. 362.

[3]. D. Evans, *Analyst, 72*, 99-102 (1947).

[4]. L. Erdey, E. Bodor, and M. Papay, *Acta Chim. Acad. Sci. Hung., 5*, 235 (1955).

[5]. L. Erdey and I. Kasa, *Talanta, 10*, 1273-1276 (1963).

[6]. L. Erdey and G. Svehla, *Z. Anal. Chem., 167*, 164 (1959).

[7]. H. Dietz and B. M. Margosches, *Chem. Ztg., 2*, 1191 (1904).

[8]. I. M. Kolthoff and R. Belcher, *Volumetric Analysis*, Vol. III, Interscience, New York, 1957, pp. 247ff.

[9]. N. H. Furman, ed., *Standard Methods of Chemical Analysis*, 6th Ed., Vol. 1, D. Van Nostrand, Princeton, N. J., 1962, pp. 518-519.

[10]. N. H. Furman, ed., *Standard Methods of Chemical Analysis*, 6th Ed., Vol. 1, D. Van Nostrand, Princeton, N. J., 1962, pp. 519-520.

[11]. N. H. Furman, ed., *Standard Methods of Chemical Analysis*, 6th Ed., Vol. 1, D. Van Nostrand, Princeton, N. J., 1962, p. 521.

[12]. K. Fajans and R. Wolff, *Z. Anorg. Allgem. Chem., 137*, 221, 245 (1924).

[13]. A. J. Berry, *Analyst, 64*, 112 (1939).

[14]. E. Schulek and E. Pungor, *Anal. Chim. Acta, 4*, 213 (1958).

[15]. G. Mannelli and M. N. Rossi, *Anal. Chim. Acta, 6*, 333 (1952).

[16]. H. V. Malmstadt and J. D. Winefordner, *Anal. Chim. Acta, 24*, 91-96 (1961).

[17]. S. A. Kiss, *Z. Anal. Chem., 207*(3), 184-186 (1965).

[18]. L. O. M. I. Smithuis and W. J. Rutten, *Pharm. Weekbl. 101*(46), 1041-1043 (1966).

[19]. S. Kotkowski and A. Lassocinska, *Chem. Anal. (Warsaw), 11*(4), 789-791 (1966).

[20]. W. Wagner, C. J. Hull, and G. E. Markle, *Advanced Analytical Chemistry*, Reinhold, N. Y., 1956, p. 235.

[21]. H. T. S. Britton, R. E. Cockaday, and J. K. Foreman, *J. Chem. Soc., 1952*, 3877.

[22]. D. D. Van Slyke, A. Hitter, and K. C. Berthelsen, *J. Biol. Chem., 74*, 659 (1927).

[23]. Z. G. Szabó and E. Sugár, *Anal. Chim. Acta, 6*, 293 (1952).

[24]. L. Erdey, Z. Bodor, and I. Buzas, *Z. Anal. Chem., 134*, 22 (1951).

[25]. J. Vulterin, *Collection Czech. Chem. Commun.* 31(6), 2501-2509 (1966).

[26]. L. Erdey, G. Svehla, and O. Weber, *Fresenius' Z. Anal. Chem. 240*(2), 91-102 (1968).

[27]. R. Belcher and A. Townshend, *Anal. Chim. Acta, 41*, 395-397 (1968)

[28]. H. Khalifa and B. Ateya, *Microchem. J., 31*(1), 147-154 (1969).

[29]. N. H. Furman, ed., *Standard Methods of Chemical Analysis*, 6th Ed., Vol. 1, D. Van Nostrand, Princeton, N. J., 1962, p. 522.

[30]. K. Gleu and W. Katthaen, *Ber. 86*, 1077 (1953).

[31]. E. Kahane, *Bull. Soc. Chim. 1948*, 70-71.

[32]. J. Vulterin, *Collection Czech. Chem. Commun., 31*(9), 3529-3535 (1966).

[33]. L. M. Likhosherstov and C. E. Brossar, *Khim. Prir. Soedin, 3*(1) 7-10 (1967).

MANGANESE

Methods for the determination of manganese include redox and complexometric titrations.

Synoptic Survey

Classical Methods

Redox Titrations. Manganese(II) is titrated with standard potassium permanganate [1].
Manganese(II) is oxidized to Mn(VII) with sodium bismuthate. The permanganate formed is back-titrated with standard sodium arsenite or ferrous sulfate [2].
High concentrations of Mn(II) may be titrated to Mn(III)

potentiometrically in pyrophosphate solution with standard potassium
permanganate [3].

Manganese(II) is oxidized to Mn(IV). The precipitate of MnO_2
so formed is dissolved in acid, potassium iodide is added, and the liberated iodine is titrated with standard thiosulfate [4]; or, after
dissolving the dioxide in sulfuric acid, excess standard ferrous sulfate is added, and the unreacted Fe(II) is back-titrated with standard
permanganate [5].

Manganese dioxide is precipitated by potassium chlorate from a
nitric acid solution. Chlorine dioxide formed is boiled off, the MnO_2
is dissolved in excess ferrous sulfate or oxalic acid, and the excess
is then titrated with standard potassium permanganate [6].

Contemporary Methods

Redox Titrations. Manganese(II) is titrated potentiometrically
in strong phosphoric acid solutions with standard potassium dichromate
[7].

Manganese(II) is oxidized to Mn(III) and titrated potentiometrically with standard solutions of iron(II), titanium(III), or chrom-
ium(II) [8].

Manganese(II) is oxidized to permanganate with silver oxide
which in turn is titrated in sulfuric acid solution with electrogenerated vanadium(IV). The end point is determined amperometrically [9].

Manganese(II) in the presence of iron(II) is titrated biamperometrically with dichromate [10].

Complexometric Titration. Manganese(II) is titrated with standard ethyleneglycolbis-(β-aminoethyl ether)-N,N-tetraacetic acid (AEGT)
using Eriochrome Black T as the indicator [11].

Outline of Recommended Contemporary Methods for Manganese

Redox Titrations

Method 1. Manganese(II) is titrated potentiometrically to Mn
(III) in a strong phosphoric acid solution (12 M) at room temperature
with standard potassium dichromate. Atmospheric oxygen does not interfere. The accuracy is 0.03% for samples containing 29-150 mg manganese
per 50 ml of solution being titrated. The end point may also be determined photometrically in which case the error is 0.3-1.0% for samples
containing 5-17 mg manganese per 40 ml of solution [7].

Method 2. Manganese(II) is oxidized by perchloric acid to a
violet manganese(III) pyrophosphate in a boiling mixture of equal
parts of 72% perchloric acid and 85% phosphoric acid. The Mn(III) thus
formed is titrated potentiometrically with standard iron(II), titan-
ium(III), or chromium(II) solutions. If titanium(III) or chromium(II)
solutions are used as the titrant, any iron present in the sample may
be titrated immediately after the manganese. Manganese(III) may also
be titrated coulometrically with electrogenerated iron(II) to a potentiometric end point. Manganese ore, Spiegeleisen, ferromanganese,
and rail steel may be analyzed by this method. The coulometric method

is the preferred one for higher precision [8].

Method 3. Samples containing manganese(II) are treated with argentic oxide to oxidize the Mn(II) to permanganate. The latter is then titrated in 0.5 F sulfuric acid with electrogenerated vanadium(IV). The end point is determined biamperometrically with two similar platinum electrodes with an applied potential of 250 mV. Chromium interferes. The use of vanadyl ion as a rather weak reducing agent makes this titration method specific for Mn(II). Precision and accuracy are good [9].

Method 4. Manganese(II) and iron(II) are titrated in 12 M phosphoric acid with a mixture of potassium dichromate and ferric ammonium sulfate dissolved in 12 M phosphoric acid and diluted to the desired concentration. Titrations are carried out under nitrogen. During the titration, changes in current are detected using platinum electrodes and a 0.5 A galvanometer. Two sharp current minima are noted which correspond to the Mn(II) and iron(II) equivalence points. The error for samples containing only manganese is 0.97-1.22%. Uranium(IV), vanadium(IV), molybdenum(V), and cerium(III) interfere. Cobalt(II), nickel(II), iron(III), tungsten(VI), uranium(VI), calcium, magnesium, and aluminum do not interfere [10].

Complexometric Titration

Samples containing micro- and semimicroamounts of manganese(II) are titrated with 0.01 M AEGT using Eriochrome Black T as the indicator. The method is simple and rapid. It compares very favorably with classical methods, and is faster than the Volhard method. Titrations are carried out at pH 9.5-10.5. High concentrations of ammonia are to be avoided otherwise manganese-ammonia complexes are formed prior to the end point and results are too low. Manganese forms a 1:1 complex with the AEGT. The error is 0.25, 0.05, and 0.33% for samples containing 0.4, 4.0, and 8.0 mg manganese respectively [11].

References

[1]. N. H. Furman, ed., *Standard Methods of Chemical Analysis*, 6th Ed., Vol. 1, D. Van Nostrand, Princeton, N. J., 1962, pp. 644-645.

[2]. T. R. Cunningham and R. W. Coztman, *Ind. Eng. Chem. 16*, 58 (1924).

[3]. J. J. Lingane and R. Karplus, *Anal. Chem.*, *18*, 191 (1946).

[4]. N. H. Furman, ed., *Standard Methods of Chemical Analysis*, 6th Ed., Vol. 1, D. Van Nostrand, Princeton, N. J., 1962, p. 651.

[5]. N. H. Furman, ed., *Standard Methods of Chemical Analysis*, 6th Ed., Vol. 1, D. Van Nostrand, Princeton, N. J., 1962, pp. 651.

[6]. N. H. Furman, ed., *Standard Methods of Chemical Analysis*, 6th Ed., Vol. 1, D. Van Nostrand, Princeton, N. J., 1962, pp. 651-652.

[7]. G. G. Rao and P. K. Rao, *Talanta*, *10*(12), 1251-1266 (1963).

[8]. J. Knoeck and H. Diehl, *Talanta*, *14*(9), 1083-1095 (1967).

[9]. D. G. Davis, *Anal. Chem.*, *31*(9), 1460-1463 (1959).

[10]. D. Singh and U. Bhatnagar, *Allg. Prakt. Chem.*, *18*(4), 107-108 (1967).

[11]. F. B. Martinez and M. P. Castro, *Anales Real Soc. Espan. Fis. y Quim. (Madrid)*, *56B*, 27-30 (1960).

TECHNETIUM

Titrimetric methods for the determination of technetium involve redox processes. Other currently available methods of analysis are radiochemical, spectrophotometric, and polarographic.

Synoptic Survey

Classical Methods

None.

Contemporary Methods

Redox Titration. Technetium(VII) is determined by coulometric reduction in an acetate buffered solution of sodium tripolyphosphate [1].

Outline of Recommended Contemporary Methods for Technetium

Redox Titration

Technetium(VII) is determined by coulometric reduction in a solution of sodium tripolyphosphate buffered at pH 4.7 with an acetate buffer. The reduction is carried out at a potential of -0.70 V versus SCE resulting in the formation of technetium(III). In the range of 0.5-5.0 mg technetium(VII), the relative error in the determination is about ± 1.0%, and the relative standard deviation is 0.5%. Accuracy and precision are better than with other currently available methods such as radiochemical, spectrophotometric, or polarographic. The following ions do not interfere: cerium(IV), cobalt(II), chromium(VI), copper(II), europium(II), manganese(VII), rhenium(VII), zinc(II), chloride, perchlorate, or sulfate. Interferences are shown by iron (III), molybdenum(VI), ruthenium(IV), uranium(VI), vanadium(IV), fluoride, and nitrate [1].

References

[1]. A. A. Terry and H. E. Zittel, *Anal. Chem.*, 35(6), 614-618 (1963).

RHENIUM

Methods for the determination of rhenium include redox and complexometric titrations.

Synoptic Survey

Classical Methods

Redox Titrations. Perrhenate is reduced with bismuth amalgam to
Re(V) followed by titration with standard ceric sulfate [1].
Rhenium is determined indirectly by precipitation as tetraphenylarsonium perrhenate which is titrated potentiometrically with standard
iodine solutions [2].

Contemporary Methods

Redox Titrations. Rhenium(VII) is titrated amperometrically with
a standard iron(II) sulfate solution in a mixture of phosphoric and
sulfuric acids [3].
Rhenium(VII) is titrated amperometrically [4], or potentiometrically [5] with standard solutions of chromium(II). The titration may
also be carried out amperometrically by titration with titanium(III)
in sulfuric acid solutions [6].
Rhenium(VII) is titrated with standard tin(II) chloride using
indigo carmine as the indicator [7].

Complexometric Titration. Rhenium(VII) is precipitated as thallous perrhenate which is dissolved in a mixture of hydrochloric acid
and bromine. The thallic ion so formed is titrated with standard EDTA
to a xylenol orange end point [8].

Outline of Recommended Contemporary Methods for Rhenium

Redox Titrations

Method 1. Rhenium(VII) is titrated amperometrically with standard Mohr's salt in a supporting electrolyte of 8 M phosphoric and 10
M sulfuric acids. Platinum or graphite electrodes are used as indicator electrodes and a SCE electrode as the reference. Iron in 500-fold
amounts, nickel and cobalt in 100-fold amounts, niobium, zirconium,
titanium, and tungsten in 10-fold amounts, or equal amounts of molyb-

denum or vanadium do not interfere. Interferences are to be noted for
chromium and manganese. The method is recommended for the analysis of
rhenium alloys [3].

Method 2. Rhenium(VII) is titrated amperometrically in a nitrogen atmosphere with standard solutions of chromium(II) at potentials
of 0.5-0.7 V. The titration is carried out in 5 M sulfuric acid as the
supporting electrolyte using rotating platinum or graphite indicator
electrodes versus SCE as the reference. Results tend to be more reproducible using the graphite electrodes because of greater ease of
cleaning. The method is used for the analysis of Mo-Re, W-Re, and Cr-
Re alloys. A 200-fold excess of nickel or cobalt, a 150-fold excess
of chromium(III), a 15-fold excess of molybdenum, a 2-fold excess of
chromium(II), or equivalent amounts of iron and tungsten do not interfere [4].

Method 3. Samples containing rhenium(VII) are titrated potentiometrically with standard solutions of chromium(II) salts. Platinum
indicator electrodes are used, and, since the potential tends to be
reestablished rather slowly after addition of portions of the titrant,
potassium iodide is added as a catalyst. Similar and reproducible results are obtained in the absence of the potassium iodide if an anodically polarized indicator electrode is employed versus a nonpolariz-
able saturated calomel electrode. The value of the potential change
at the equivalence point during the reduction of the rhenium(VII) to
rhenium(IV) is about 380 mV. The indicator electrode may be anodically
polarized with currents of $2.5 \times 10^{-5} - 6.25 \times 10^{-7}$ A/cm^2. The accuracy in the determination is 0.3% [5].

Method 4. Samples containing rhenium(VII) may be titrated amperometrically in 5 M sulfuric acid with standard titanium(III) salt
solutions. The titration is best carried out under a nitrogen atmosphere using rotating platinum or graphite electrodes at 0.8 V versus
SCE. Up to 200-fold amounts of cobalt or nickel, 100-fold amounts of
iron or molybdenum or a 0.75-fold amount of tungsten do not interfere
[6].

Method 5. Samples containing 3-52 mg rhenium(VII) in 5 M hydrochloric acid are titrated with 0.04 M tin(II) chloride prepared by
dissolving the required amount of the tin(II) chloride in 6 M hydrochloric acid. During the titration which is performed under a carbon
dioxide atmosphere, the rhenium(VII) is reduced to rhenium(V). The
end point is detected using the blue to yellow-green color change of
indigo carmine. The mean standard deviation is 0.05 mg rhenium. The
use of the redox indicator greatly simplifies the usual procedure [7].

Complexometric Titration

Rhenium(VII) is precipitated quantitatively from neutral or
weakly acidic solutions as thallous perrhenate. This precipitate of
TlReO$_4$ is separated and dissolved in a mixture of hydrochloric acid
and bromine. The bromine oxidizes the thallous ion to thallic ion
which is then titrated at pH 5 with standard EDTA using xylenol orange
as the indicator. Chlorate and perchlorate do not interfere. Molyb-

date, permanganate, thiocyanate, bromate, iodate, and the halides must be absent. As little as 1 mg of rhenium can be determined within ± 1.0% relative error [8].

References

[1]. H. Sputzy, R. J. Magee, and C. L. Wilson, *Mikrochim. Acta, 1957*, 354.

[2]. H. H. Willard and G. Smith, *Ind. Eng. Chem. Anal. Ed., 11*, 186 (1939).

[3]. Z. A. Gallai and T. Ya. Rubinskaya, *Zh. Analit. Khim. 22*(9), 1378-1381 (1967). Eng. transl. *J. Anal. Chem. USSR, 22*(9), 1161-1164 (1967).

[4]. Z. A. Gallai, T. Ya. Rubinskaya, and A. V. Fursova, *Zh. Analit. Khim., 21*(5), 584-589 (1966). Eng. transl. *J. Anal. Chem. USSR, 21*(5), 522-526 (1966).

[5]. V. A. Zarinskii and V. A. Frolkina, *Zh. Analit. Khim., 17*(11), 75-79 (1962).

[6]. Z. A. Gallai and T. Ya. Rubinskaya, *Zh. Analit. Khim., 21*(8), 961-964 (1966). Eng. transl. *J. Anal. Chem. USSR, 21*(8), 857-860 (1966).

[7]. G. Henze and R. Geyer, *Z. Anal. Chem., 200*(6), 434-438 (1964).

[8]. H. Hamaguchi and R. Sugisitz, *Bunseki Kagaku, 10*, 1256-1258 (1961).

GROUP VIII

Iron	Ruthenium	Osmium
Cobalt	Rhodium	Iridium
Nickel	Palladium	Platinum

IRON

Because of the widespread distribution in nature of iron in both divalent and trivalent states, the determination of iron is represented by a voluminous literature encompassing a wide spectrum of procedures. This section will be devoted to a summary of classical and contemporary methods for the determination of ferrous and ferric ion, and of ferrocyanide and ferricyanide ion by titrimetric methods. Such determinations involve primarily redox, complexometric, and precipitation reactions.

Synoptic Survey of Ferrous Ion and Ferrocyanide Ion Determinations

Classical Methods

Redox Titration. Many standard oxidizing agents have been used for the titration of ferrous ion. Titrations may involve the following titrants: potassium dichromate in acid solution using barium diphenylamine sulfonate as the indicator [1]; potassium permanganate in acid solution [2]; ceric sulfate using o-phenanthroline as the indicator [3]; and potassium bromate either electrolytically [4] or with auric ion as the indicator [5].

Contemporary Methods

Redox Titrations. Iron(II) is titrated potentiometrically in 10 M phosphoric acid with standard potassium bromate [6].

Iron(II) is titrated amperometrically or spectrophotometrically with cerium(IV) using ferroin as the indicator [7,8].

Ferrous ion may be determined in the presence of vanadium(IV) by a differential procedure involving a titration with ceric ion followed by a titration of the vanadium(V) formed with standard ferrous ammonium sulfate [9].

Ferrocyanide is titrated potentiometrically with standard permanganate or cerium(IV) in acid solution [10].

Ferrocyanide is determined by direct coulometry [11,12].

Iron(II) in a solution of methyl cellosolve and carbon tetra-

196

chloride is titrated with standard antimony pentachloride solution to
a congo red end point [13].

Iron(II) is titrated potentiometrically with standard ceric sulfate, dichromate or bromate using 4'-hydroxy-N,N-dimethyl-4-amino-azo-
benzene as the indicator [14].

Iron(II) and cobalt(II) may be determined by successive amperometric titrations with standard ammonium vanadate [15].

Iron(II) and manganese(II) are determined in a mixture by titration with standard dichromate. A biamperometric end-point detection
is employed [16].

Iron(II) is determined by titration with standard vanadate solutions using promethazine as the redox indicator [17].

Outline of Recommended Contemporary Methods for Ferrous Ion

Redox Titrations

Method 1. Iron(II) is titrated potentiometrically in 10 M phosphoric acid with standard potassium bromate in an inert atmosphere at
room temperature. Various organic substances such as oxalate, tartrate, citrate, sucrose, or glucose do not interfere. The method is
advocated for the determination of iron in pharmaceutical preparations.
The average error is 0.2% [6].

Method 2. Samples containing iron(II) in concentrations of about
1×10^{-5} M in dilute sulfuric acid solution are titrated amperometrically with standard cerium(IV) at 0.65 V versus SCE using a rotating
platinum wire electrode and ferrous phenanthroline as the indicator.
The end point may also be determined spectrophotometrically at 500 mμ
if ferroin is used as the indicator. In this case, the titration is
best performed at 0-5°. Precision and accuracy are better than 1%
[7,8].

Method 3. In a mixture of iron(II) and vanadium(IV), the sum of
the components is measured by titration in sulfuric acid with standard
cerium(IV) sulfate using methyl orange as the indicator. This indicator is destructively oxidized by the titrant at the end point. Phosphoric acid is then added, and the vanadium(V) formed in the first
titration is titrated reductometrically to vanadium(IV) with standard
ferrous ammonium sulfate using barium diphenylamine sulfonate as the
redox indicator. Precision and accuracy are good [9].

Method 4. Samples containing ferrocyanide are titrated potentiometrically with standard potassium permanganate or cerium(IV) sulfate in acid solution. Best results are obtained in 1 N sulfuric
acid. The following bimetallic electrode pairs can be used: platinum-tungsten, platinum-palladium, platinum-silver, platinum-nickel,
or platinum-gold. From 20-100 mg of potassium ferrocyanide contained
in 50 ml can be determined [10].

Method 5. Ferrocyanide is determined by oxidation coulometry.
Good results are possible if the potential of the platinum disc
electrode (anode) is controlled at + 0.55 - + 0.85 V versus SCE in the

presence of 0.1 M potassium chloride as the supporting electrolyte.
Amounts ranging from 50-150 mg per 100 ml can be determined with a
reproducibility of $\pm$ 1%. Bromide ion does not interfere. Iodide and
thiocyanate interfere. Cyanide ion does not interfere if the determination is carried out below + 0.70 V versus SCE. Errors range from
2-0.05 mg at levels of 150-50 mg ferrocyanide respectively. The method
is more direct and less time consuming than other presently available
titrimetric and spectrophotometric methods [11,12].

Method 6. Samples containing the equivalent of 0.1-0.2 M of
iron(III) in 10 ml of methyl cellosolve are reduced in a carbon dioxide
atmosphere with about 200 gm of zinc amalgam in the presence of carbon
tetrachloride. After about fifteen minutes reaction time, the mixture
is filtered and the filtrate is titrated with antimony pentachloride
solution using congo red as the indicator. Results are good. The
procedure has also been applied to the determination of tin(IV) and
titanium(IV) after reduction to the lower valence state. The tin(II)
so formed is titrated to a methyl orange end point while the titanium
(III) is titrated to a safranin T end point [13].

Method 7. Samples containing ferrous ion are titrated with standard solutions of cerium sulfate, dichromate, or bromate using 4'-
hydroxy-N,N-dimethyl-4-aminoazobenzene as the indicator. A two electron mechanism is involved in the oxidation of the indicator whose redox potential is 0.918 V versus the standard hydrogen electrode. The
standard deviation ranges from $\pm$ 0.05 to $\pm$ 0.07 for the above three
oxidizing agents [14].

Method 8. Successive amperometric titrations with ammonium vanadate are used to determine iron(II) and cobalt(II) in the same solution. Two platinum electrodes are used at a constant voltage of 50
or 100 mV. The sample is first titrated in 2 N sulfuric acid at 50
mV to determine the iron. EDTA is then added to mask the iron(III).
The acidity is adjusted to 0.5 N, the voltage is raised to 100 mV, and
the titration is continued with the same titrant. Equivalence points
are determined graphically. Excellent agreement is found between the
amounts of iron and cobalt taken and the amounts determined [15].

Method 9. Manganese(II) and iron(II) are determined in the same
solution by biamperometric titrimetry. Titrations are carried out
under nitrogen in 12 M phosphoric acid. Two sharp minima in the titration curve are noted for the two equivalence points. Results are
good. Uranium(IV), vanadium(IV), molybdenum(V) and cerium(III) interfere since they are oxidized under the same conditions. Cobalt(II),
nickel(II), iron(III), tungsten(VI), uranium(VI), calcium, magnesium,
and aluminum do not interfere [16].

Method 10. Samples containing iron(II) are titrated with sodium
vanadate using promethazine hydrochloride, 10-(2-dimethylamino-1-
propyl)phenothiazine hydrochloride as the redox indicator. The latter
undergoes a one electron oxidation to a red intermediate which undergoes a further one electron oxidation to a colorless sulfoxide. The
titration is carried out in either 1 M sulfuric acid or 1 M hydrochloric acid containing a few milliliters of 85% phosphoric acid using

as the titrant 0.05 N sodium vanadate. The indicator color change is from blue-green to violet at the end point. For best results at least 0.7 ml of a 0.1% solution of the indicator in a total volume of 75 ml must be used. Results are good [17].

Synoptic Survey of Ferric Ion and Ferricyanide Ion Determinations

Classical Methods

Redox Titrations. Ferric ion is titrated with standard titanous chloride or sulfate in acid solution. The end point may be detected electrometrically or by means of thiocyanate as a color indicator [18, 19].

Iron(III) is reduced with iodide ion and the iodine so formed is titrated with standard thiosulfate [20].

Ferric ion is titrated potentiometrically with standard chromium (II) solutions [21].

Iron(III) may be titrated with stannous chloride with a potentiometric detection of the end point [22].

Complexometric Titration. Samples containing ferric ion are titrated with standard EDTA using tiron or salicylic acid as the indicator. The end point may also be detected spectrophotometrically [23,24].

Precipitation Titrations. Ferric oxinate is determined by titration with standard bromate [25].

Samples containing iron(III) may be determined by high-frequency titration with oxine solution [26].

Contemporary Methods

Redox Titrations. Ferric ion is titrated coulometrically with electrogenerated chlorocuprous ion in a hydrochloric-sulfuric acid medium. The end point is detected potentiometrically [27].

Microgram and milligram amounts of iron(III) are determined by controlled-potential coulometry [28].

Ferricyanide is titrated in basic solution with standard vanadium (II) using diphenyldianisidine-o,o'-carboxylate, barium diphenylamine sulfonate, or methylene blue as the indicator [29].

Iron(III) is determined by titration with standard copper(I) chloride in aqueous hydrochloric acid [30].

Iron(III) is titrated with electrogenerated titanium(III) coulometrically with bioamperometric end-point detection [31].

Complexometric Titrations. Solutions of iron(III) are treated with an excess of equimolar cerium(III)-EDTA. The cerium(III) released is titrated with standard EDTA to a xylenol orange end point [32,33].

In a mixture of iron(II) and iron(III), the iron(III) is first determined by titration with standard EDTA at pH 4.7-5.0. Excess EDTA is then added, the pH is raised to pH 6.7-7.2, and the iron(II)-EDTA is titrated with standard ferricyanide [34].

Samples containing iron(III) are titrated thermochemically with standard sodium fluoride [35].

Outline of Recommended Contemporary Methods for Ferric Ion and
Ferricyanide

Redox Titrations

   Method 1. Samples of iron containing 5-50 mg as ferric ion are
titrated coulometrically with electrogenerated chlorocuprous ion in
hydrochloric-sulfuric acid medium. The end point is detected potentiometrically. Up to a chloride ion concentration of about 0.5 M, the
approximate course of the reaction is:

$$FeCl^{+2} + CuCl_2^- = Fe^{+2} + Cu^{+2} + 3 Cl^-$$

The mean error for the above sample size is $+ 0.0001$ mg and the average
deviation is ± 0.018 mg. By reversing the polarity of the platinum
generator electrode, iron(II) can be titrated anodically with electrogenerated chlorine and, therefore, the method may be adapted to the
analysis of mixtures of iron(II) and iron(III). Only a few metals
interfere [27].

   Method 2. Microgram and milligram amounts of iron(III) are determined by controlled-potential coulometry forming iron(II). The determination is carried out in either 1 N sulfuric or 1 N hydrochloric
acids. For the analysis of silicates, nonferrous and ferrous metals,
or alloys, sulfuric acid is to be preferred. Hydrochloric acid is used
for those samples which contain a high calcium content such as fluor-
spar, dolomite, limestone, and magnesite. The main interference,
copper, is readily removed by electrodeposition. The error in the
determination of milligram amounts of iron is less than 0.5% [28].

   Method 3. Ferricyanide ion is titrated in a basic solution with
standard vanadium(II) sulfate. Several indicators are suitable such
as diphenyl dianisidine-o,o'-dicarboxylate, barium diphenylamine sulfonate, or methylene blue. In acid medium there is a two electron
transfer which serves as the basis for this potentiometric titration
[29].

$$V^{+2} + 2 Fe(CN)_6^{-3} = V^{+4} + 2 Fe(CN)_6^{-4}$$

   Method 4. Samples containing 0.68-2.9% iron are determined by
titration with a standard solution of copper(I) chloride in aqueous
hydrochloric acid. The end point may be detected visually or potentiometrically. The solution of the titrant is stable for long periods
of time, and since the redox potential of the copper(II)/copper(I)
system is low, it may be used for the quantitative determination of
other oxidants such as vanadate or chlorate [30].

   Method 5. The titanium(III) generated quantitatively at a copper
or copper-amalgam cathode in 2 N hydrochloric acid in the presence of
added thiocyanate ion is used for the determination of iron(III). The
titanium(III) is generated at a potential more negative than $+ 0.3$ V
versus SCE and the determination is carried out coulometrically with
biamperometric end-point detection. For samples containing 0.5-1.0
mg iron in 5 ml, the error is 0.5% [31].

Complexometric Titrations

Method 1. The two serious limitations in the use of the classical titration of iron(II) with a standard solution of a strong oxidizing agent are, first of all, interferences from common metals such as
chromium, copper, molybdenum, mercury, titanium, tungsten, and vanadium; and, second, in the analysis of alloys high in chromium, copper,
and nickel, the iron must be separated prior to its determination preferably on an ion exchange column and eluted with acetone. The deter-
minative step proper, however, can be carried out via an elegant sub-
stitution titration. The iron is first separated by a column extraction procedure, eluted with acetone and an excess of equimolar cerium
(III)-EDTA is added. The cerium(III) released by the iron(III) is then
titrated with standard EDTA to a xylenol orange end point. The iron
(III)-EDTA complex (log K_s = 24) is more stable than the cerium(III)-
EDTA complex (log K_s = 16). Thus the reaction

$$Fe^{+3} + CeY^{-1} = Ce^{+3} + FeY^{-1}$$

proceeds rapidly and quantitatively. The relative standard deviation
for the procedure is 0.01% [32,33].

Method 2. Both of the EDTA complexes of iron(III), log K_s = 25.1,
and of iron(II), log K_s = 14.3, are quite stable, and, as a result,
titration of a mixture containing iron(II) and iron(III) allows one to
determine each constituent. Samples containing iron(II) and iron(III)
containing 0.00625-0.00125 mole of each are acidified with hydrochloric
acid followed by the addition of glycine or sodium acetate to raise
the pH to 4.8-5.0. A voltage of 30 mV is applied to a pair of platinum
electrodes placed in the solution. Titration is carried out with 0.1
M EDTA. The slope of the current-voltage polarization curves at the
null potential is used for end point detection. The square wave technique is used to measure the slope. The iron(III) is titrated first
with the standard EDTA. After the iron(III) end point, enough 0.1 M
EDTA is added to complex all the iron(II) and to provide an excess of
about 10 ml. The pH is then raised to 6.6-7.2 by the addition of sodium hydroxide, and the titration is continued with 0.1 M ferricyanide
to titrate the iron(II)-EDTA complex. Both end points are V-shaped.
Precision and accuracy are very good [34].

Method 3. Samples containing iron(III) are determined by thermo-
chemical titration with 0.84-0.1 N sodium fluoride. Good end points
are obtained in the linear temperature-volume of titrant curves for
the determination of iron, aluminum, copper(II), and lead. In the
case of iron, the complex formed is Na_3FeF_6. Precision is about 1%
[35].

References

[1]. N. H. Furman, ed., *Standard Methods of Chemical Analysis*, 6th
     Ed., Vol. 1, D. Van Nostrand, Princeton, N. J., 1962, pp. 542-
     544.

[2].   N. H. Furman, ed., *Standard Methods of Chemical Analysis*, 6th
       Ed., Vol. 1, D. Van Nostrand, Princeton, N. J., 1962, pp. 544-
       547.

[3].   G. F. Smith, *Cerate Oxidimetry*, G. F. Smith Chemical Company,
       Columbus, Ohio, 1964.

[4].   I. M. Kolthoff and J. J. Vleeschhouwer, *Rev. Trav. Chim., 45,*
       923 (1926).

[5].   L. Szebellédy and V. Madis, *Z. Anal. Chem., 114,* 249 (1938).

[6].   G. G. Rao and N. K. Murty, *Z. Anal. Chem., 208*(2), 97-101 (1965).

[7].   I. M. Kolthoff and R. Woods, *Microchem. J., 5*(4), 569-572 (1961).

[8].   I. M. Kolthoff and B. B. Bhayia, *Microchem. J., 4,* 451-457 (1960).

[9].   K. B. Rao, R. I. Joseph, and O. Devaraj, *Chemist-Analyst, 55*(4),
       103-104 (1966).

[10]. K. Ueno and T. Tachikawa, *Nippon Kagaku Zasshi, 82,* 577-580
       (1961).

[11]. G. Sivaramaiah and V. R. Krishnan, *Z. Anal. Chem., 231*(3), 192-
       194 (1967).

[12]. G. Sivaramaiah and V. R. Krishnan, *Z. Anal. Chem., 206*(1), 33-
       35 (1964).

[13]. C. Yoshimura, H. Hara, and T. Hara, *Bunseki Kagaku, 16*(9), 883-
       887 (1967).

[14]. V. Matrka, J. Sadila, V. Sporek, and J. Pipalova, *Collection
       Czech. Chem. Commun., 32*(1), 344-351 (1967).

[15]. D. Singh and U. Bhatnagar, *J. Indian Chem. Soc., 44*(3), 237-
       238 (1967).

[16]. D. Singh and U. Bhatnagar, *Allg. Prakt. Chem., 18*(4), 107-108
       (1967).

[17]. H. S. Gowda, R. Shakunthala, and G. Ramappa, *Talanta, 15*(2),
       266-269 (1968).

[18]. W. M. Thornton, Jr., and R. Roseman, *Am. J. Sci.,* (5), *20,* 14
       (1930).

[19]. R. A. Chalmers, in *Comprehensive Analytical Chemistry*, C. L.
       Wilson and D. W. Wilson, eds., Elsevier, New York, 1962, pp.
       635-655.

[20]. N. H. Furman, ed., *Standard Methods of Chemical Analysis*, 6th
       Ed., Vol. 1, D. Van Nostrand, Princeton, N. J., 1962, p. 551.

[21]. J. J. Lingane, *Anal. Chem.*, *20*, 797-799 (1948).

[22]. E. Müller and J. Gorne, *Z. Anal. Chem.*, *73*, 385 (1928).

[23]. P. B. Sweetser and C. E. Bricker, *Anal. Chem.*, *25*, 253 (1953).

[24]. E. G. Brown, *Metallurgia*, *49*, 101, 151 (1954).

[25]. I. Kitijima, *J. Chem. Soc. Japan*, *55*, 884 (1934).

[26]. G. Goto and T. Hirayama, *J. Chem. Soc. Japan*, *73*, 656 (1952).

[27]. J. J. Lingane, *Anal. Chem.*, *38*(11), 1489-1494 (1966).

[28]. G. W. C. Milner and J. W. Edwards, *Analyst*, *87*, 125-133 (1962).

[29]. J. P. Tandon and K. L. Chawla, *Bull. Chem. Soc. Japan*, *40*(4), 992-993 (1967).

[30]. L. A. Kastakina and M. Ya. Shapiro, *Izv. Vyssh. Ucheb. Zaved. Khim. Khim Tekhnol.*, *11*(1), 14-16 (1968).

[31]. Z. Slovak and M. Přibyl, *Fresenius' Z. Anal. Chem.*, *228*(4), 266-275 (1967).

[32]. S. S. Yamamura, *Anal. Chem.*, *36*(9), 1858-1861 (1967).

[33]. J. S. Fritz and C. E. Hedrick, *Anal. Chem.*, *34*, 1411 (1962).

[34]. L. C. Hall and D. A. Flanigan, *Anal. Chem.*, *33*, 1495-1498 (1961).

[35]. P. Deschamps, A. Debruck, and Y. Bonnaire, *Anal. Chim. Acta*, *40*(2), 259-267 (1968).

COBALT

Titrimetric methods available for the determination of cobalt involve primarily redox and complexometric procedures. However, none of the methods is absolutely specific for cobalt and, hence, some type of preliminary separation is usually required.

Synoptic Survey

Classical Methods

Redox Titrations. Cobalt(II) is titrated potentiometrically with standard potassium ferricyanide in an ammoniacal solution [1], or ethylene diamine [2].

Cobalt(III) from the oxidation of cobalt(II) by hydrogen per-oxide is treated with iodide ion and the liberated iodine is titrated with standard thiosulfate [3].

Complexometric Titration. In the absence of nickel, cobalt(II) is titrated with standard potassium cyanide [4].

Contemporary Methods

Redox Titrations. Cobalt(II) is titrated potentiometrically with standard potassium ferricyanide [5].
Cobalt(II) is titrated amperometrically with ferricyanide in an aqueous ammonium citrate or aqueous glycine solution [6].
Submilligram amounts of cobalt(II) are determined by titration with ferricyanide. The end point is detected photometrically [7].
Cobalt(III) is determined by sample treatment with excess $K_4Fe(CN)_6$ followed by back titration with standard ceric sulfate [8].

Complexometric Titrations. In samples containing nickel, the nickel is removed by precipitation with dimethylglyoxime, the cobalt (II) is isolated by ion exchange, and finally titrated with EDTA to an eriochrome T end point [9,10].
Microamounts of cobalt are determined by titration of the blue complex formed between acetone and cobalt(II) with standard EDTA [11].
Cobalt(II) is determined indirectly via its pyridine thiocyanate complex [12].

Outline of Recommended Contemporary Methods for Cobalt

Redox Titrations

Method 1. The classical titration of cobalt(II) with standard potassium ferricyanide is improved by application of constant current potentiometric end-point detection. The break in the titration curve at the end point is sharp, and there is no longer need to plot titration curves. The potential change at the equivalence point is about 0.5 V. In the actual analysis, an excess of 0.1 M ferricyanide is added to the cobalt sample in a solution buffered with ammonium hydroxide and ammonium citrate. The unreacted ferricyanide is back titrated with standard cobalt sulfate. The standard deviation for 8.26 meq cobalt is ± 0.04. In the presence of 0.025 meq manganese, the standard deviation is ± 0.11 [5].

Method 2. Samples containing 0.2-30.0 mg cobalt(II) are titrated amperometrically with standard potassium ferricyanide in aqueous ammonium citrate or aqueous glycine solution at pH 9.8 or 8.0, respectively. In ammonium citrate medium, the titrations are carried out at potentials of -0.30 V and -1.55 V [lying on the plateaus of the polarograms of 0.005 M cobalt(II) and 0.005 M ferricyanide, respectively]. In glycine buffer, however, the titration is successful only at a potential on the ferricyanide wave at -0.55 V. In citrate solution, cerium(III) and iron(II) interfere, whereas, in glycine solutions, interferences are met in the cases of copper(II) and vanadium(V). The error in the titration of 10^{-5} M cobalt solution varies between 1.0 and 1.8% [6].

Method 3. Submilligram amounts of cobalt(II) cannot be titrated

potentiometrically in ammoniacal solution with ferricyanide because of
the extremely sluggish response of the platinum electrode. The products
of the reaction

$$Co^{2+} + Fe(CN)_6^{-3} = Fe(CN)_6^{-4} + Co^{3+}$$

exhibit together a much higher optical density than would be expected
from straight forward addition considerations. The reason is not known.
Furthermore, the effect is not sufficiently reproducible to be utilized
for the spectrophotometric analysis of cobalt, but does not prevent the
use of photometric end-point detection. The titration is feasible at
a cobalt(II) concentration of 10^{-4} M with a light path length of 4 cm.
The method allows the detection of 0.06 mg cobalt in a volume of 10 ml.
The error is about 1%. Up to 120 mg iron, 90 mg nickel, 60 mg copper,
or 30 mg chromium(III) does not interfere. No interference is met from
a mixture of 15 mg chromium(III), 15 mg iron(III), 15 mg nickel, and
15 mg copper. The method is very useful for the rapid determination
of small amounts of cobalt [7].

Method 4. Samples containing cobalt(III) are determined by adding
an excess of standard potassium ferrocyanide and back-titrating the
excess with standard ceric sulfate. Silver, calcium, strontium, barium,
nickel, lead, copper, zirconium, and thorium interfere. Sodium, potassium, lithium, magnesium, cadmium, and zinc do not interfere. With
slight modifications, the method is also applicable to the determination
of iridium(III) in which case the interfering ions are lithium, zinc,
cobalt, cadmium, indium, gold, ruthenium, platinum, uranium, and tungsten [8].

Complexometric Titrations

Method 1. Large amounts of cobalt are determined by a modified
procedure of Lewis and Straub [9]. This modification consists in the
back-titration of the excess added standard EDTA with standard mangan-
ese(II) using eriochrome T as the indicator. The results of the titration are much improved by careful control of the conditions of the titration such as pH, temperature, volume of the solution, the concentration of the reducing agent, and the rate at which the titration is
performed. The precision as calculated by 2σ (reproducibility at the
95% confidence limits) is $\pm$ 0.07 and $\pm$ 0.20 for samples containing 8%
and 43% cobalt respectively [9,10].

Method 2. A rapid and accurate titration method for the determination of microamounts of cobalt(II) is based on the titration of
the blue solution which results upon the addition of acetone to a
neutral or slightly acidic (acetic acid) solution of cobalt(II) containing an excess of ammonium thiocyanate. This blue solution is titrated with standard EDTA to a colorless or very slightly pink end
point. As little as 2.5 γ/ml of cobalt can be determined visually and
1 γ/ml photometrically. Pyrophosphate, citrate, oxalate, and cations,
except for the alkali metals, interfere. In the presence of nickel,
cobalt may be extracted with amyl alcohol while the interferences caused
by copper, aluminum, and the alkaline earths are best removed chemically prior to the titration [11].

Method 3. Samples containing 5-25 mg cobalt(II) are treated with 0.2 N thiocyanate, and with 0.5 ml pyridine. The mixture is diluted to about 25 ml warmed a few minutes and filtered. An aliquot is taken, the pH is adjusted to about 3.1 with formic acid, and the unreacted thiocyanate ion is titrated with 0.1 N mercuric nitrate in the presence of diphenyl carbazone as the indicator. Iron(III), chromium (III), aluminum(III), bismuth(III), antimony(III), tin(II), and tin (IV) do not interfere. For a sample containing 4.425 mg cobalt, the amount of cobalt found was 4.418 mg. This method is also applicable to the determination of copper, cadmium, nickel, zinc, and manganese [12].

References

[1]. N. H. Furman, ed., *Standard Methods of Chemical Analysis*, 6th Ed., Vol. 1, D. Van Nostrand, Princeton, N. J., 1962, pp. 387-388.

[2]. H. Diehl and J. P. Butler, *Anal. Chem.*, *27*, 777 (1955).

[3]. H. A. Laitinen and L. W. Burdett, *Anal. Chem.*, *23*, 1268 (1951).

[4]. N. H. Furman, ed., *Standard Methods of Chemical Analysis*, 6th Ed., Vol. 1, D. Van Nostrand, Princeton, N. J., 1962, pp. 386-387.

[5]. B. Kratochvil, *Anal. Chem.*, *35*(9), 1313-1314 (1963).

[6]. W. U. Malik and H. Om, *Talanta*, *14*(11), 1341-1343 (1967).

[7]. H. Poppe and G. den Boef, *Talanta*, *10*(12), 1297-1298 (1963).

[8]. O. C. Saxena, *Microchem. J.*, *13*(1), 120-124 (1968).

[9]. L. L. Lewis and W. A. Straub, *Anal. Chem.*, *32*, 96 (1960).

[10]. G. A. Bauer, *Anal. Chem.*, *36*(3), 691-693 (1964).

[11]. R. N. Sarma, *Anal. Chem.*, *32*(6), 717-719 (1960).

[12]. A. L. J. Rao and B. K. Puri, *Fresenius' Z. Anal. Chem.*, *235*(2), 176-178 (1968).

NICKEL

Methods for the determination of nickel include indirect acid-base titrations, indirect redox titrations, and complexometric titrations.

Synoptic Survey

Classical Methods

Acid-Base Titration. Nickel is precipitated with dimethylglyoxime. The precipitate is dissolved in an excess of standard sulfuric acid, and the excess is titrated with standard base [1,2,3].

Redox Titration. Nickel is precipitated as the glyoxamate. The precipitate is hydrolyzed in acidic solution and the hydroxylamine formed is oxidized with standard ferric ion solution. The resulting ferrous ion is then determined by titration with standard permanganate [4], or the liberated hydroxylamine is oxidized by the addition of a known excess of standard potassium bromate, and the unreacted bromate is back titrated with standard arsenite [5].

Complexometric Titrations. Nickel samples are titrated with standard potassium cyanide solution to form the complex $K_2Ni(CN)_4$. Silver iodide is added before the titration and its disappearance indicates the end point [6].
Nickel is precipitated with dimethylglyoxime. The precipitate is dissolved and titrated with standard EDTA using murexide [7,8] or alizarin S as the indicator [9].
Nickel may be determined in the presence of cobalt by adding a known excess of EDTA and back titrating this excess with standard magnesium solutions. This first titration gives the sum of both the cobalt and nickel. The titration mixture is then treated with hydrogen peroxide and potassium cyanide. The EDTA liberated by this treatment (proportional to the amount of nickel present) is then titrated with standard magnesium sulfate solution [10].

Contemporary Methods

Redox Titrations. Samples containing trivalent nickel are dissolved in acidic solutions of Mohr's salt. The iron(III) formed is complexed with phosphoric acid, and the excess iron(II) is titrated with standard dichromate [11].
Nickel samples are dissolved and the nickel is plated out on a platinum electrode. The nickel is displaced by an equivalent of silver by dipping in a silver nitrate solution. The silver is then dissolved from the electrode with nitric acid and precipitated as silver iodide. Finally the iodide is oxidized to iodate and determined iodometrically with standard thiosulfate [12].

Complexometric Titrations. Samples of nickel are titrated potentiometrically with standard tetraethylenepentamine (Tetren) [13].
Nickel ion is titrated at pH 8.2 with standard nitrilotriacetic acid (NTA) using murexide as the indicator [14].
Nickel ion is determined in the presence of cobalt by first masking the cobalt with potassium cyanide and hydrogen peroxide. Upon the addition of silver nitrate, an equivalent amount of nickel is displaced and is titrated with standard EDTA [15].
Nickel samples are determined by treatment with excess thiocyanate in pyridine solutions. The excess thiocyanate not involved in complex

formation is then titrated with standard mercury(II) solutions using diphenylcarbazone as the indicator [16].

Nickel is determined by titration with EDTA using potassium thiocarbonate as the metal indicator [17].

Nickel samples in an ammoniacal buffer are treated with a solution of Hg-EDTA plus a few drops of disodium-EDTA and then titrated potentiometrically with standard copper sulfate using a mercury electrode as the indicator electrode [18].

Outline of Recommended Contemporary Methods for Nickel

Redox Titrations

Method 1. Samples of NiO containing nickel(III) and weighing about 0.5 gm are dissolved in an excess of 8% hydrochloric acid which is 0.02 M in Mohr's salt. The dissolution is carried out under a carbon dioxide atmosphere and requires about ten minutes. Phosphoric acid is then added to complex the iron(III), and the excess Mohr's salt is titrated with 0.02 M potassium dichromate using diphenyl amine as the indicator. The error is $\pm$ 0.01-0.02% for samples containing 0.02-0.50% nickel [11].

Method 2. Nickel samples are analyzed by an indirect iodometric procedure. The method is also applicable to the determination of copper and cobalt. The sample containing the nickel is dissolved, and the nickel is plated out quantitatively on a platinum electrode. The electrode is then placed in a silver nitrate solution. The displacement which occurs replaces the nickel with an equivalent amount of silver. The electrode now carrying its coating of silver is treated with nitric acid, and the dissolved silver is precipitated as the iodide. Finally, the iodide is oxidized to iodate and determined iodometrically by titration with standard thiosulfate using a starch indicator. For samples containing 4.35 mg nickel the error is less than 1% [12].

Complexometric Titrations

Method 1. Samples containing nickel are titrated potentiometrically with 0.01 M Tetren. The end point is established with a mercury indicator and a calomel reference electrode. The alkaline earths, aluminum, rare earths, bismuth, lead, or scandium do not interfere. The mean deviation for a sample containing 9 mg nickel is -0.2% [13].

Method 2. Nickel is titrated at pH 8.2 with standard NTA using murexide as the indicator. The reaction proceeds very rapidly. Magnesium and calcium do not interfere. During the titration, the pH is adjusted periodically using an ammonium hydroxide-ammonium chloride buffer. At the end point the color change of the indicator is from yellow to reddish violet [14].

Method 3. Samples containing up to 50 mg nickel are determined in the presence of cobalt by masking both ions with potassium cyanide and hydrogen peroxide in an ammoniacal solution. Ten percent aqueous silver nitrate is then added until a slight turbidity develops in the

solution. By this treatment, the nickel is selectively demasked. The
nickel is then titrated immediately with 0.05 M EDTA using murexide as
the indicator. The determination may also be carried out indirectly
by adding an excess of EDTA to the solution after demasking the nickel.
The excess EDTA is then titrated with calcium solution using Methylthymol Blue as the indicator. Precision and accuracy are very good
even for samples containing up to 30 mg of cobalt per 50 mg nickel [15].

Method 4. Samples containing 5-25 mg nickel are determined via
their pyridine thiocyanate complexes. The dissolved sample usually as
the nitrate is added to a known excess of 0.2 N potassium thiocyanate.
A small quantity of pyridine is then added, the mixture is diluted
with water, heated, and the complex is removed by filtration. The pH
of an aliquot of the filtrate is adjusted to about 3.1 with dilute
formic acid, and the excess thiocyanate ion is titrated with 0.1 N
mercury(II) nitrate using diphenylcarbazone as the indicator. Accuracy and precision are excellent. The method is also applicable to the
determination of copper, cadmium, cobalt, zinc, and manganese [16].

Method 5. Samples containing 5-20 mg nickel are determined by
titration in a solution buffered with an ammonium buffer to pH 10-11.
The titrant is 0.02 M EDTA and potassium thiocarbonate, K_2CS_3, is used
as the metal indicator. Nickel forms a water soluble blood-red thiocarbonate complex whose dissociation constant is 1 x 10^{-9}. The complex is stable only at high pH values and is decomposed on acidification. The titration is carried out by adding to the sample the buffer
and 2-3 ml of 1% aqueous potassium thiocarbonate. Titration of the
red solution is then carried out with the EDTA to a yellowish green
color. The average error is about 1%. All metal ions yielding precipitates with thiocarbonate or whose hydroxides precipitate interfere.
Among the anions, phosphate and oxalate interfere [17].

Method 6. Samples of iron-nickel alloy containing 0.5-30 γ
nickel are dissolved in hydrochloric acid, evaporated to dryness, redissolved in dilute nitric acid, and the iron precipitated and removed
as the hydroxide. The sample is then treated with an anion exchange
resin to remove chloride ion. To a 2-4 ml aliquot of the solution so
prepared there is added an ammonium buffer, 0.05 ml of 1 x 10^{-3} M Hg-
EDTA, a few drops of 0.01 M disodium-EDTA and the solution is then
titrated potentiometrically with 0.005 M copper sulfate using a mercury microelectrode as the indicator. A potential change of "several
dozen" millivolts is noted at the equivalence point. The maximum
error is 2-3% for the above sample size [18].

References

[1]. N. H. Furman, ed., *Standard Methods of Chemical Analysis*, 6th
Ed., Vol. 1, D. Van Nostrand, Princeton, N. J., 1962, p. 702.

[2]. S. W. Parr and J. M. Lingren, *Trans, Am. Brass Founders' Assoc.*,
5, 120 (1912).

[3].   J. Holluta, *Monatsh.*, *40*, 281 (1919).

[4].   M. B. Tourgrinoff, *Ann. Soc. Sci. Bruxelles*, *548*, 314 (1934).

[5],   N. H. Furman and J. F. Flagg, *Anal. Chem.*, *12*, 738 (1940).

[6],   N. H. Furman, ed., *Standard Methods of Chemical Analysis*, 6th
        Ed., Vol. 1, D. Van Nostrand, Princeton, N. J., 1962, p. 707.

[7].   W. F. Harris and J. R. Sweet, *Anal. Chem.*, *24*, 1062 (1952).

[8].   W. F. Harris and J. R. Sweet, *Anal. Chem.*, *26*, 1648 (1954).

[9].   K. Ter Haar and J. Bazen, *Anal. Chim. Acta*, *14*, 413 (1956).

[10]. R. Přibil, *Chem. Age (London)*, *72*, 141 (1955).

[11]. A. G. Chervinko, P. M. Mel'nik, and S. S. Lisnyak, *Zh. Anal.
        Khim.*, *22*(9), 1428-1429 (1967). Eng. transl. *J. Anal. Chem.
        USSR*, *22*(9), 1205-1206 (1967).

[12]. H. Weiz and M. Gonner, *Microchim. Acta*, *1968*, 355-357.

[13]. C. N. Reilly and A. Vavoulis, *Anal. Chem.*, *31*, 243-248 (1959).

[14]. G. Numajiri, M. Kodama, and A. Shimizu, *Nippon Kagaku Zasshi*,
        81, 454-457 (1960).

[15]. R. Přibil and V. Vesely, *Talanta*, *13*(3), 515-518 (1966).

[16]. A. L. J. Rao and B. K. Puri, *Fresenius' Z. Anal. Chem.*, *235*(2)
        176-178 (1968).

[17]. K. N. Johri and K. Singh, *Bull. Chem. Soc. Jap.*, *40*(4), 990-
        992 (1967).

[18]. I. P. Alimarin and M. N. Petrikova, *Zh. Anal. Khim.*, *21*(10)
        1257-1258 (1966). Engl. transl. *J. Anal. Chem. USSR*, *21*(10),
        1114-1115 (1966).

RUTHENIUM

Only a few methods for the titrimetric determination of ruthenium appear in the literature. Except possibly for one report, all
of the methods require prior isolation of ruthenium. All reported
determinations involve redox titration [1,2,3].

Synoptic Survey

Classical Methods

Redox Titrations. Ruthenium is reduced by an excess of standard tin(II) chloride which is then back-titrated with iodine to a starch end point [4].

Ruthenium(IV) is titrated potentiometrically to ruthenium(III) with standard titanium(III) chloride [5,6].

Contemporary Methods

Redox Titrations. Ruthenium(IV) is titrated potentiometrically to ruthenium(VII) with standard lead tetraacetate [7].

Ruthenium(IV) is titrated with electrogenerated titanium(III) with the equivalence point being determined potentiometrically, amperometrically, or photometrically [8,9].

Outline of Recommended Contemporary Methods for Ruthenium

Redox Titrations

Method 1. Samples containing 5-16 mg ruthenium in the tetravalent state are oxidized to ruthenium(VII) by titration with 0.1 N lead tetraacetate. The titration is carried out in 0.03-0.1 N perchloric acid and the end point is determined potentiometrically. If 0.05 N tetraacetate is used as the titrant, 2-10 mg ruthenium can be determined in a volume of 50 ml. The method is probably as precise as any available at the present time. Solutions containing ruthenium in a valence state below four are oxidized prior to titration with chlorine water in 6 N hydrochloric acid. The mean deviation for samples containing 2-10 mg ruthenium is of the order of 0.05% [7].

Method 2. Ruthenium is determined by titration with electrogenerated titanium(III). The electrogeneration of the titanium (100% current efficiency) can be best carried out at a current density of 1-5 mg/cm^2 on a tungsten or graphite electrode at 20° in a medium of $\geq$ 9 N hydrochloric acid $\geq$ 8 N sulfuric acid, or at 50° in a medium of $\geq$ 6 N hydrochloric acid or $\geq$ 2 N sulfuric acid. Graphite is somewhat less efficient than tungsten. The end point is detected potentiometrically at zero current, amperometrically with two polarized electrodes, or photometrically. This procedure can also be used for the determination of iridium(IV). A photometric determination of the end point is recommended for ruthenium. The sensitivity is 0.4 γ ruthenium per milliliter. Osmium, rhenium and ruthenium(II) do not interfere. Platinum(IV) interferes [8,9].

References

[1]. F. E. Beamish, *The Analytical Chemistry of the Noble Metals, International Series of Monographs in Analytical Chemistry* (R. Belcher and L. Gordon, ed., Vol. 1, Pergamon Press, New York, 1966.

[2]. W. P. Griffith, *The Chemistry of the Rarer Platinum Metals, (Os, Ru, Ir, and Rh).*, Interscience, New York, 1967.

[3]. F. E. Beamish, *Talanta, 13*(8), 1953-1968 (1966).

[4]. J. L. Howe, *J. Am. Chem. Soc., 49*, 2393-2395 (1928).

[5]. W. R. Chrowell and D. M. Yost, *J. Am. Chem. Soc., 50*, 374-381 (1928).

[6]. N. K. Pshenitsyn and S. I. Ginzburg, *Izvest. Sektora Platiny i Drug. Blagorod. Metal Inst. Obshchei i Neorg. Khim., Akad. Nauk S.S.S.R., 32*, 20-30 (1955).

[7]. I. Nemec, A. Berka, and J. Zyka, *Microchem. J., 6*, 525-537 (1962).

[8]. N. I. Stenina and P. K. Agasyan, *Zh. Analit. Khim., 21*(8), 965-969 (1966). Eng. transl. *J. Anal. Chem. USSR, 21*(8), 861-864 (1966).

[9]. N. I. Stenina, P. K. Agasyan, and G. A. Brentsveig, *Zh. Analit. Khim. 22*(1), 91-95 (1967). Eng. transl. *J. Anal. Chem. USSR, 22*(1), 74-78 (1967).

RHODIUM

Methods for the determination of rhodium are not numerous. The reported titrations involved are either redox or precipitation titration [1].

Synoptic Survey

Classical Methods

Redox Titration. Samples of rhodium(III) are oxidized to rhodium(V) with sodium bismuthate and titrated with Mohr's salt solution using phenylanthranilic acid as the indicator [2].

Precipitation Titrations. Rhodium is precipitated by the addition of an excess of standard thionalide solution and the excess is back-titrated with standard iodine solution to a slight excess of iodine and then again back-titrated with standard thiosulfate [3].

Rhodium samples are treated with stannous chloride and titrated with a standard solution of sodium 1-piperidine carbodithionate [4].

Contemporary Method

Redox Titration. Rhodium is precipitated as the hydroxide, $Rh(OH)_3$, oxidized to RhO_2 with hypochlorite and the latter is determined iodometrically [5].

Outline of the Contemporary Method for Rhodium

Redox Titration

Rhodium present in phosphate plating baths is precipitated with a base as $Rh(OH)_3$. Sodium hypochlorite is added and the rhodium(III) is oxidized to RhO_2. After removal of the excess oxidizing agent by boiling the solution, the rhodium is determined iodometrically.

$$2 RhO_2 + 2 I^- + 8 H^+ + 12 Cl^- = 2 RhCl_6^{3-}$$

$$+ I_2 + 4 H_2O$$

The liberated iodine is titrated immediately with 0.05 N sodium thiosulfate. Iron, nickel, cobalt, copper, lead, and manganese interfere. Results agree within 3% with those obtained by gravimetric analysis [5].

References

[1].   F. E. Beamish, *The Analytical Chemistry of the Noble Metals, International Series of Monographs in Analytical Chemistry,* (R. Belcher and L. Gordon, ed.,) Vol. 24, Pergamon, New York, 1966.

[2].   V. S. Syrokomskii and N. N. Proshenkova, *Zh. Analit. Khim. 2,* 247-252 (1947).

[3].   H. Kienitz and L. Rombock, *Z. Anal. Chem., 117,* 241 (1939).

[4].   N. K. Pshenitsyn and N. V. Fedorenko, *Zh. Analit. Khim., 44,* 588-589 (1959).

[5].   H. Rackett, *Metal Finishing, 59,* 79-80 (1961).

PALLADIUM

Methods for the determination of palladium involve redox, complexometric, and precipitation titrations [1].

Synoptic Survey

Classical Methods

Redox Titration. Palladium as the chloropalladate(IV) is reduced with excess Mohr's salt, and the unreacted ferrous ion is back-titrated with standard potassium chromate [2].

Precipitation Titrations. Palladium solutions are titrated with standard solutions of nickel diethyldithiophosphate [3].
Palladium in alloys is determined by titration with potassium iodide [4,5].

Contemporary Methods

Redox Titration. Palladium(II) is titrated amperometrically to palladium(IV) with standard hypochlorite [6].

Complexometric Titrations. Palladium(II) is titrated amperometrically with standard 2,4-dithiobiuret [7].

Milligram amounts of palladium are determined by the addition of excess standard EDTA and back-titrating the excess with mercuric nitrate [8].

Palladium(II) is determined by amperometric titration with benzimidazol-2-yl-methanthiol [9].

Palladium is titrated amperometrically with guanylthiourea [10], or with Reinecke's salt [11].

Precipitation Titrations. Palladium in the presence of gold and platinum is determined by potentiometric titration with standard potassium iodide [12].

Palladium in the presence of silver is determined by titration with potassium iodide. The end point is determined potentiometrically or with an external starch-iodate indicator [13].

Outline of Recommended Contemporary Methods for Palladium

Redox Titration

Up to now no oxidimetric procedure has been successful in making use of the half cell reaction $Pd(II) \rightarrow Pd(IV) + 2 e$. This is due to the fact that no medium has been investigated in which there is a marked difference in the oxidation potentials of the palladium and the oxidant. Recently it has been found, however, that palladium(II) can be titrated amperometrically with standard hypochlorite in a 0.5 M chloride-0.05 M azide supporting electrolyte at pH 2. Success in the titration is apparently due to the stabilizing influence of this supporting electrolyte on the palladium(IV) species accompanied at the same time by a marked reduction in the oxidation potential of the Pd(IV)/Pd(II) couple. The end point is found with a rotating platinum electrode at a potential of 0.5 V versus SCE. Large amounts of various metals and oxygen may be present; only a few substances interfere. Furthermore, if necessary, and if desired, the azide can be easily removed before the determination of other ions. Samples containing 0.5-2.1 mg palladium can be titrated with a precision of about 0.5% [6].

Complexometric Titrations

Method 1. Palladium(II) forms soluble 1:2 complexes with 2,4-dithiobiuret in >1 M hydrochloric, sulfuric, or nitric acid and 2:3 complexes at pH 3-7. Samples containing 26-1000 γ palladium are titrated amperometrically with 0.001 M 2,4-dithiobiuret in 0.1-1.0 M hydrochloric acid at 0.5 V versus SCE. Cadmium, zinc, lead, bismuth, tin, nickel, and cobalt do not interfere. No interference is met with mercury and silver in 1.0 M hydrochloric acid. Iron(III), iridium(IV), and gold(III) interfere due to their oxidizing effect on the titrant. Iron(III) is masked with fluoride, iridium(IV), and gold(III) are reduced with ferrous ion. If present, platinum can be masked with a large amount of ammonium chloride. Fiftyfold amounts of antimony(III) and mercury(II) as well as fivefold to thirtyfold amounts of arsenic (III) and tellurium(II) reduce the diffusion current but do not shift the end point. Only selenium interferes. All values obtained by this method agree closely with those of the gravimetric method [7].

Method 2. Samples containing 0.1-4 mg palladium are determined by the addition of excess standard EDTA and back-titrating the unreacted EDTA with standard mercuric nitrate solutions in a basic medium. The method is applicable in the presence of iron(III), indium, lanthanum, and thorium. Accuracy is very good [8].

Method 3. Milligram amounts of palladium are titrated amperometrically with benzimidazol-2-yl-methanthiol. The titration is carried out in an acetate buffer at pH 4-5 at an applied potential of -0.2 V versus SCE. Mercury(I), mercury(II), titanium(IV), and copper interfere. The method may also be applied to the determination by amperometric titration of both silver and copper [9].

Method 4. Samples of palladium containing 0.1-1.5 mg/50 ml are titrated at 0.95-1.0 V versus SCE with 0.01 M guanylthiourea. The titration may be carried out over a wide pH interval. The ratio of Pd:guanylthiourea depends on the pH. At pH values greater than pH 6, the Pd:guanylthiourea ratio is 1:1. For the above sample sizes, the relative error is less than 1%. Gold(III) and iron(III) interfere and must be removed or reduced with ferrous sulfate and masked with fluoride ion [10].

Method 5. Samples containing 0.05-13.5 mg palladium are determined amperometrically by a procedure based on titration with Reinecke's salt. A rotating platinum wire electrode is used as the indicator electrode and the titration is carried out in 0.5 M hydrochloric acid at + 0.9 V versus SCE at 800 rpm. Solutions of hydrochloric acid and sulfuric acid are both recommended as the best supporting electrolytes. Interferences from gold(III), antimony(III), tin(II), copper(II), and bismuth are minimized by masking with pyrophosphate and citrate. The relative error for samples containing less than 2.90 mg palladium is less than 1%. Calcium, strontium, barium, magnesium, cobalt, nickel, iron(III), iron(II), chromium(III), tin(II), manganese(II), selenium (IV), and tellurium(IV) do not interfere when present in a ratio of 1:80 [11].

Precipitation Titrations

Method 1. Palladium is determined in the presence of platinum by potentiometric titration of an aliquot containing 5-70 mg with standard potassium iodide solution. A gold or palladium indicator electrode is used together with a SCE reference. Platinum is determined in another aliquot containing 3-30 mg platinum by potentiometric titration with standard copper(I) chloride. The same electrode system is used as in the titration of palladium. The mean square error for the determination of samples containing 3-20 mg palladium in the presence of platinum is 0.05 mg, and for the determination of platinum (3-20 mg) in the same sample, the mean square error is 0.06 mg. When the palladium content of the sample is 10-20 mg, 10-50 mg copper, large amounts of nickel, iron up to 50 mg [provided that sodium fluoride is added to complex the iron(III)], and up to 10 mg tellurium do not interfere. Interferences are met with in the case of selenium. In samples containing gold and palladium, both are easily determined in the same solution. The gold content is determined first by titration with standard hydroquinone solution in a solution which is 0.6 N in sulfuric acid. Alternately, the titration of the gold may be carried out with standard ascorbic acid in 0.10-0.13 N sulfuric acid. Titrations for gold are carried out at 60-70°. Then, after adjusting the acidity to 0.15-0.3 N sulfuric acid, the palladium is titrated with standard potassium iodide [12].

Method 2. Palladium(II) and silver(I) form complexes with ammonia having the composition $Pd(NH_3)_4^{2+}$ and $Ag(NH_3)_2^+$, respectively. Upon decreasing the pH of a solution containing these two complex ions, the dissociation of the complexes is increased.

$$Pd(NH_3)_4^{2+} + 4 H^+ = Pd^{2+} + 4 NH_4^+ \qquad (1)$$

$$Ag(NH_3)_2^+ + 2 H^+ = Ag^+ + 2 NH_4^+ \qquad (2)$$

At pH values lower than 7, the equilibrium of Eq (1) is shifted to the left while that of Eq (2) is moved towards the right. If, for analysis, the solution is adjusted to pH 4-5 and then titrated with iodide, silver iodide is precipitated while the palladium remains in solution until all of the silver ions have been removed. The first drop of iodide solution in excess of that equivalent to the silver present causes the precipitation of brown palladium(II) iodide which changes the pale yellow turbidity of the solution to grey-brown. The method is well suited for the determination of samples containing 3-30 mg silver and 7-70 mg palladium in 10-20 ml of solution. The end point for the silver is determined by the internal indicator effect in which the yellow turbidity changes to a grey-brown. The end point in the titration of the palladium is determined by means of an external starch-iodate indicator or potentiometrically using an iodine electrode [13].

References

[1]. F. E. Beamish, *The Analytical Chemistry of the Noble Metals, The International Series of Monographs in Analytical Chemistry* (R. Belcher and L. Gordon, eds., Vol. 24, Pergamon Press, New York, 1966.

[2]. V. S. Syrokomskii and S. M. Gubel'bank, *Zh. Anal. Khim.*, *4*, 146 (1949).

[3]. A. I. Busev and M. I. Ivanyutin, *Zh. Anal. Khim.*, *13*, 18-30 (1958).

[4]. B. H. Atkinson, R. N. Rhoda, and R. G. Lowell, *Analyst*, *79*, 368-370 (1954).

[5]. R. N. Rhoda and R. H. Atkinson, *Anal. Chem.*, *28*, 535-537 (1956).

[6]. R. G. Clem and E. H. Huffman, *Anal. Chem.*, *40*(6), 945-948 (1968).

[7]. A. S. Sukhoruchkina and Yu. I. Usatenke, *Tr. Dnepropetr. Khim.-Tekhnol. Inst. 1963*(16), 35-42.

[8]. H. Khalifa and M. M. Khater, *Z. Anal. Chem.*, *191*, 339-345 (1962).

[9]. M. M. Chakrabartly, *Talanta*, *13*(8), 1186-1190 (1966).

[10]. A. S. Sukhoruchkina and Yu. I. Usatenko, *Khim. Tekhnol. Respub. Mezhvedom. Nauch.-Tekh. Sb. 1967*, No. *7*, 81-86.

[11]. I. L. Bagbanly and B. Z. Rzaev, *Dokl. Akad. Nauk. Azerb. USSR.*, *23*(9), 19-22 (1967).

[12]. N. K. Psenitsyn, S. I. Ginzburg, and I. V. Prokof'eva, *Zh. Anal. Khim.*, *7*(3), 343-346 (1962).

[13]. G. A. Segar, *Analyst*, *87*, 230-232 (1962).

OSMIUM

For the noble metal osmium, the only titrimetric methods reported in the chemical literature involve redox titration [1].

Synoptic Survey

Classical Methods

Redox Titrations. Osmium distillates in sodium hydroxide are acidified with sulfuric acid. The osmium is then reduced with bismuth granules and to it is added a known excess of standard metavanadate. The unreacted metavanadate is back-titrated with iron(II) solution [2].

Osmium tetroxide samples are treated with potassium iodide in sulfuric acid and the liberated iodine is titrated with thiosulfate [3,4].

The potassium iodide titration of osmium may also involve the detection of the end point potentiometrically [5].

Osmium is titrated potentiometrically with hydrazine sulfate [4], titanium(III) chloride [6], chromium(III) sulfate [7], or thiosulfate [8].

Contemporary Methods

Redox Titrations. Osmium(VII) is reduced to osmium(IV) and titrated potentiometrically with potassium permanganate in the presence of 3% telluric acid [9].

Osmium samples are reduced with bismuth and the osmium(IV) is titrated potentiometrically with cerium(IV) sulfate, potassium permanganate, or lead tetraacetate [10].

Solutions of osmium tetroxide are titrated with titanium(III) chloride using 1-hydroxy-7-aminophenoxazin-3-one or 1-hydroxy-7-dimethylaminophenoxazin-3-one as the indicator [11].

Microamounts of osmium are determined by an indirect titration with potassium ferrocyanide, followed by titration of the excess with cerium(IV) sulfate using N-phenyl-anthranilic acid as the indicator [12].

Outline of Recommended Contemporary Methods for Osmium

Redox Titrations

Method 1. Osmium(IV) is oxidized to osmium(VIII) by permanganate in basic solution. In the presence of telluric acid the following reactions occur:

$$OsCl_6^{2-} + 4 H_2O = OsO_2 \cdot H_2O + 6 Cl^- + 4 H^+ \tag{1}$$

$$OsO_2 \cdot H_2O + 6 H^+ + TeO_6^{6-} + 4 OH^- \tag{2}$$

$$= [OsO_2(OH)_4]^{2-} + TeO_3^{2-} + 5 H_2O$$

$$3 TeO_3^{2-} + 2 MnO_4^- + H_2O = 3[HTeO_6]^{2-} \tag{3}$$

$$+ 2 MnO_2 + 2 OH^-$$

$$3[OsO_2(OH)_4]^{2-} + 2 MnO_4^- = 3 OsO_5^{2-} + 2 MnO_2 \tag{4}$$

$$+ 2 OH^- + 5 H_2O$$

In the presence of the telluric acid, the manganese(IV) oxide formed is dissolved giving a red colored solution. In the actual analysis, the osmium(IV) is titrated potentiometrically with standard potassium permanganate in the presence of three percent telluric acid. The standard deviation for 8 mg osmium is -1.2%, for 19 mg is -0.4%, and for 37 mg is +0.21% [9].

Method 2. Samples containing 0.5-10.0 mg osmium in 0.4 N sodium hydroxide are acidified with 6 N sulfuric acid and passed through a reductor charged with 0.2-0.75 mm diameter bismuth granules. The reductor is rinsed with 6 N sulfuric acid and the combined solution and washes are titrated potentiometrically with 0.01 N cerium(IV) sulfate, potassium permanganate, or lead tetraacetate using a platinum indicator electrode. The inflection point is ca. 950 mV versus SCE. The error in the determination is less than $\pm$ 1.5% [10].

Method 3. Solutions of osmium tetroxide are titrated with 0.01 N titanium(III) chloride using 1-hydroxy-7-aminophenoxazin-3-one or 1-hydroxy-7-dimethylaminophenoxazin-3-one as the indicator. The relative error for samples containing 4 mg osmium is = 0.57%. During the titration, the osmium(VIII) is reduced to osmium(IV) and at the end point the color change of the indicator is from red to yellow-green. Titrations are carried out under a carbon dioxide atmosphere in the presence of ca. 2 N hydrochloric acid. The procedure can also be used for the determination of iron(III), gold(III), ferricyanide, dichromate, and vanadate [11].

Method 4. Microamounts of osmium are determined by an indirect titration. A known excess of 0.0176 N potassium ferrocyanide is added to an aqueous solution of osmium tetroxide which has been acidified with 0.1 N hydrochloric acid. The excess unreacted ferrocyanide is then titrated with 0.01 N cerium(IV) sulfate dissolved in 8 N sulfuric acid using N-phenylanthranilic acid as the indicator. The osmium tetroxide reacts with the ferrocyanide forming a green paramagnetic complex having the composition:

$$Os(VIII)O_5Os(VI)[Fe(CN)_6]_2^{2-}$$

Silver, thallium(I), cobalt(II), nickel, lead, palladium(II), zinc, bismuth, ruthenium(III), rhodium(III), gold(III), lanthanum(III), and platinum(IV) interfere [12].

References

[1]. F. E. Beamish, *The Analytical Chemistry of the Noble Metals, The International Series of Monographs in Analytical Chemistry,* (R. Belcher and L. Gordon, eds.), Vol. 24, Pergamon Press, New York, 1966.

[2]. V. S. Syrokomskij, *Compt. Rend. Acad. Sci. USSR, 46,* 280 (1945).

[3]. E. A. Klobbie, *Kon. Akad. Wetensch. Amsterdam; abstracted in J. Chem. Soc., 1899,* 76ii, 184.

[4]. W. R. Cromwell and H. D. Kirschman, *J. Am. Chem. Soc., 51,* 175-179 (1929).

[5]. D. I. Ryabchikov, *J. Applied Chem. USSR, 17,* 326-328 (1944).

[6]. W. R. Cromwell and H. D. Kirschman, *J. Am. Chem. Soc., 51,* 1695-1702 (1929).

[7]. W. R. Cromwell and H. L. Baumbach, *J. Am. Chem. Soc., 57,* 2607-2609 (1935).

[8].   D. I. Ryabchikov, *Zh. Analit. Khim.*, *1*, 47-56 (1946).

[9].   J. Santrucek, I. Nemec, and J. Zyka, *Mikrochim. Acta*, *1966* (1-2), 10-16.

[10]. J. Santrucek, I. Nemec, and J. Zyka, *Czech. Chem. Commun.* *3*(6), 2679-2688 (1966).

[11]. E. Ruzicka, *Mikrochim. Acta*, *1967*(2), 277-286.

[12]. O. C. Saxena, *Microchem. J.*, *12*(4), 609-611 (1967).

IRIDIUM

Several methods are recorded in the literature for the analysis of iridium. These involve primarily redox titration in which iridium (III) is oxidized, or iridium(IV) is reduced [1].

Synoptic Survey

Classical Methods

Iridium(IV) is titrated potentiometrically with standard solutions of hydroquinone [2,3].
Solutions of iridium(III) are oxidized with cerium(IV) sulfate and then titrated reductometrically with iron(II) sulfate or Mohr's salt solutions [4].

Contemporary Methods

Iridium(IV) in the presence of rhodium(III) is titrated with electrogenerated ferrocyanide. The end point is detected potentiometrically [5].
Iridium(IV) is titrated potentiometrically or photometrically in the presence of rhodium(III) with electrogenerated iron(II) [5].
Small amounts of iridium(IV) are titrated coulometrically in the presence of rhodium(III) with electrogenerated copper(I) [6,7].
Small amounts of iridium(IV) are titrated amperometrically with standard hydroquinone or ascorbic acid [8,9].
Microamounts of iridium(IV) are determined by amperometric titration with electrogenerated titanium(III) [10].
Iridium(IV) is determined by the addition of a known excess of standard potassium ferrocyanide followed by back-titration of the excess with ceric(IV) sulfate [11].

Outline of Recommended Contemporary Methods for Iridium

Redox Titrations

Method 1. Iridium is titrated in the presence of rhodium(III)

with electrogenerated ferrocyanide ion. The end point is detected potentiometrically. The ferrocyanide is generated from ferricyanide in
sulfuric acid solution of pH 2. The presence of eightfold amounts of
rhodium does not interfere. Precision and accuracy are good [5].

Method 2. Iridium is titrated with electrogenerated iron(II)
in the presence of rhodium(III). The end point is detected potentiometrically or photometrically. The presence of rhodium in up to sixfold
amounts does not affect the determination. The iron(II) is generated
from a solution of ferric ammonium sulfate using about a 1000-fold excess of the latter with respect to iridium. A small amount of phosphoric acid and sufficient sulfuric acid is added to make its final
concentration 4 N. The percent error averages about 2% [5].

Method 3. Small amounts of iridium are determined by coulometric
titration with copper(I) electrogenerated from cupric sulfate in 4 M
hydrochloric acid using platinum or graphite electrodes. The end point
is determined potentiometrically or amperometrically using two polarized platinum electrodes. A tenfold excess of osmium(IV), rhodium(III),
or palladium(II) does not interfere. The sensitivity is about 10^{-5} M
or approximately 2 µg/ml. Iron(III) interferes at all concentrations.
Precision and accuracy are good. The following advantages are claimed:
generation of the copper(I) with 100% current efficiency; standardiza-
tion of the titrant is not necessary; purification of the solutions is
not necessary; and, finally, titration in an inert gas atmosphere is
not required since the iridium(IV) values are not affected by air [6,
7].

Method 4. Samples containing 10^{-2} to 10^{-5} M iridium are titrated amperometrically with standard hydroquinone or ascorbic acid solutions. Rhodium, platinum, copper, nickel, lead, selenium, or tellurium do not interfere. If phosphoric acid is added to mask the iron,
up to 100-fold excess iron(III) may be present. The relative error
in the titration with hydroquinone ranges from 2-8% for samples containing between 8 and 0.05 mg iridium, respectively. In the titrations
with ascorbic acid the relative error varies from 1-7% for samples containing between 9.4 and 0.03 mg iridium respectively [8,9].

Method 5. Small amounts of iridium(IV) are determined by amperometric titration with electrogenerated titanium(III). The end point
is measured using two polarized platinum electrodes. A 100% current
generating efficiency for titanium(III) at a current density of 1-5
mA/cm^2 is obtained by using a tungsten electrode at 20° in a medium
of $\geq$ 9 N hydrochloric acid or $\geq$ 8 N sulfuric acid or at 50° in a medium $\geq$ 6 N hydrochloric acid or $\geq$ 2 N sulfuric acid. The method is
applicable in the presence of 10-20-fold amounts of osmium(IV), rhen-
ium(IV) or palladium(II). Platinum(IV) interferes. The sensitivity
is about 0.2 µg/ml [10].

Method 6. Samples containing 2.3-17 mg/liter iridium(IV) are
determined by the addition of a known excess of 0.0176 N potassium
ferrocyanide. After standing for twenty-four hours the excess ferrocyanide is back-titrated with 0.01 N cerium(IV) sulfate in 8 N sulfuric
acid; N-phenylanthranilic acid is used as the indicator. Lithium,

zinc, cobalt, cadmium, indium, gold, ruthenium, platinum, uranium, and tungsten interfere. No interferences are met with in the case of sodium, potassium, barium, calcium, strontium, magnesium, lanthanum, or thorium [11].

References

[1]. F. E. Beamish, *The Analytical Chemistry of the Noble Metals, The International Series of Monographs in Analytical Chemistry,* (R. Belcher and L. Gordon, eds.), Vol. 24, Pergamon, New York, 1966.

[2]. D. I. Ryabchekov, *J. Applied Chem. USSR, 17,* 326-328 (1944).

[3]. W. B. Pollard, *Bull. Inst. Mining Met. 497,* 9-17 (1948).

[4]. W. A. E. McBride and M. C. Cluett, *Can. J. Research, 28B,* 788-798 (1950).

[5]. N. I. Stenina and P. K. Agasyan, *Zh. Anal. Khim.,* 20(3), 351-354 (1965). Eng. transl. *J. Anal. Chem. USSR,* 20(3), 322-324 (1965).

[6]. N. I. Stenina and P. K. Agasyan, *Zh. Anal. Khim.,* 20(2), 196-199 (1965). Eng. transl. *J. Anal. Chem. USSR,* 20(2), 177-180 (1965).

[7]. N. I. Stenina and P. K. Agasyan, *Zh. Anal. Khim.,* 21(10), 1223-1226 (1966). Eng. transl. *J. Anal. Chem. USSR,* 21(10), 1084-1086 (1966).

[8]. N. K. Pshenitsyn and N. A. Ezerskaya, *Zh. Anal. Khim.,* 14, 81-86 (1959).

[9]. D. B. Labkovskaya and R. I. Andreeva, *Analiz. Blagorod. Met. Moscow, 1965,* 41-46.

[10]. N. I. Stenina, P. K. Agasyan, and G. A. Berentsveig, *Zh. Anal. Khim.,* 22(1), 91-95 (1967). Eng. transl. *J. Anal. Chem. USSR,* 22(1), 74-78 (1967).

[11]. O. C. Saxena, *Microchem. J.,* 13(1), 120-124 (1968).

PLATINUM

Most of the titrimetric methods for the determination of platinum which appear in the literature cannot be highly recommended. Those which do appear involve primarily redox, complexometric, or precipitation titration [1].

Synoptic Survey

Classical Methods

Redox Titrations. Platinum(IV) is reduced to platinum(II) with cuprous chloride and the chloro complex, $PtCl_4^{2-}$, is titrated with standard potassium permanganate to $PtCl_4^{2-}$ [2,3].

Platinum(IV) is reduced with excess iron(II) and the excess is back-titrated with standard ammonium metavanadate [4].

Platinum(IV) is determined after reduction to platinum(II) with hydrazine by titration with permanganate or bromate [5].

Platinum is determined by reduction to platinum(II) with ascorbic acid and subsequently titrated potentiometrically with ferric chloride [6].

Complexometric Titration. In the presence of gold, platinum is titrated with standard sodium diethyldithiocarbamate [7].

Precipitation Titration. Platinum is precipitated as the chloroplatinate, reduced to the metal, and the released chloride is titrated with silver nitrate [8].

Contemporary Methods

Redox Titrations. Platinum in the presence of palladium is titrated potentiometrically with standard copper(I) chloride [9].

Platinum(IV) is determined indirectly by precipitation with excess standard ferrocyanide followed by titration of the unreacted ferrocyanide with cerium(IV) sulfate [10].

Complexometric Titrations. Platinum(II) is titrated by the constant current coulometric generation of a sulfhydryl group from monothioethylene glycol (MTEG) [11,12].

Platinum(IV) is determined by addition of a known excess of EDTA and subsequently the excess is back-titrated with standard zinc or lead acetate [13].

Outline of Recommended Contemporary Methods for Platinum

Redox Titrations

Method 1. Samples containing 3-30 mg platinum in the presence of 5-70 mg palladium are titrated potentiometrically with standard copper(I) chloride. In the presence of gold(III) the direct potentiometric titration of platinum(IV) is not feasible. In this case only one common potential jump is observed during the titration with standard copper(I) chloride which corresponds to the reduction of both platinum(IV) to platinum(II) and of gold(III) to the metallic state. In the presence of gold, however, it is possible to calculate the platinum value from two titrations. This is accomplished by determining the sum of gold and platinum by a potentiometric titration. The amount of gold then can be determined in another aliquot by titration with standard ascorbic acid. During the titration with ascorbic acid, the gold is reduced from gold(III) to gold(II). Nevertheless, values are acceptable only if the platinum content is considerably greater than that of gold. It is also possible to determine platinum, palladium, and gold in the same mixture if base metals are absent and if the platinum and palladium concentrations are considerably higher than those of gold. Titration of two different aliquots in a carbon dioxide atmosphere are required. In the first aliquot, the gold(III) is first determined by titration with hydroquinone in 0.6 M sulfuric acid medium, or by titration with ascorbic acid in 0.1-0.13 M sulfuric acid medium either titration being carried out at 60-70°. Then using the same solution, the acidity is adjusted to 0.15-0.3 N sulfuric acid, and the palladium is titrated with standard potassium iodide at room temperature. In the second aliquot, having an acidity of 0.1-0.5 N sulfuric acid, the total gold plus platinum is determined by a potentiometric titration with cuprous chloride at room temperature. Platinum values are obtained by difference [9].

Method 2. Samples of platinum(IV) are determined indirectly by precipitation of the platinum with a known excess of standard potassium ferrocyanide. Subsequently, the excess is titrated with cerium(IV) sulfate in the presence of 2 M sulfuric acid. N-phenylanthranilic acid is used as the indicator. For samples containing 5-82 mg platinum, the accuracy is within $\pm$ 0.2 mg. Rhodium, ruthenium, gold, and platinum(II) interfere [10].

Complexometric Titrations

Method 1. Platinum(II) is titrated with the mercury(II) complex of monothioethylene glycol which has sufficient water solubility to permit constant current coulometric generation of its sulfhydryl group. The end point in the titration is determined potentiometrically. In the case of this titrant, interferences from secondary complexes involving the solubilizing hydroxyl group are minimal. The direct titration of platinum(II) is slow. It is better to generate a slight excess of sulfhydryl and then to back-titrate after about one minute. The accuracy is about $\pm$ 1% [11,12].

Method 2. Samples containing 5-30 mg platinum(IV) as the chloro complex $PtCl_6^{2-}$ in dilute hydrochloric acid buffered with an acetate buffer to pH 5.5 are treated with a twofold to threefold excess of EDTA. A 1:1 complex is formed between the platinum and the EDTA. The excess EDTA is finally titrated with standard 0.02 N zinc acetate or lead acetate. The average error in the determination is $\pm$ 0.1%. A slight modification in the procedure permits one to determine palladium in the same sample. In the case of palladium, the error is $\pm$ 0.2% [13].

References

[1]. F. E. Beamish, *The Analytical Chemistry of the Noble Metals*, *The International Series of Monographs in Analytical Chemistry*, (R. Belcher and L. Gordon, eds.), Vol. 24, Pergamon, New York,

1966, pp. 343-348.

[2]. A. A. Grinberg and Z. E. Goldbraikh, *J. Gen. Chem. USSR*, *14*, 808 (1944).

[3]. A. A. Grinberg and A. I. Vobroborskaya, *Zh. Neorg. Khim.*, *1*, 2360-2367 (1956).

[4]. V. S. Syrokomskii and N. N. Prokof'eva, *Zh. Anal. Khim.*, *1*, 83 (1946).

[5]. O. Stelling, *Svensk. Kem. Tidskr.*, *43*, 130 (1931).

[6]. E. A. Maksimyuk, *Izvest. Sektora Platiny Y Drug. Blagorod. Metal.*, *Inst. Obshchei i Neorg. Khim.*, *Akad. Nauk SSSR 30*, 180-182 (1955).

[7]. W. B. Pollard, *Trans. Inst. Mining Met.*, *47*, 331-346 (1937-1938).

[8]. E. Hintz, *Z. Anal. Chem.*, *35*, 72-73 (1896).

[9]. N. K. Pshenitsyn, S. I. Ginzburg, and I. V. Prokof'eva, *Zh. Anal. Khim.*, *17*, 343-346 (1962).

[10]. O. C. Saxena, *Talanta*, *13*(6), 862-863 (1966).

[11]. B. Miller and D. N. Hume, *Anal. Chem.*, *32*(7), 764-767 (1960).

[12]. B. Miller and D. N. Hume, *Anal. Chem.*, *32*(7), 524-528 (1960).

[13]. P. G. Shakova, *Zavodsk. Lab.*, *32*(10), 1201-1202 (1966).